Matteo Carrara

Side reactions in diesel particulate filters

Matteo Carrara

Side reactions in diesel particulate filters

formation of high mutagenic nitrated polycyclic aromatic hydrocarbons

Südwestdeutscher Verlag für Hochschulschriften

Publisher:
Südwestdeutscher Verlag für Hochschulschriften
is a trademark of
Dodo Books Indian Ocean Ltd., member of the OmniScriptum S.R.L Publishing group
str. A.Russo 15, of. 61, Chisinau-2068, Republic of Moldova Europe
Printed at: see last page
ISBN: 978-3-8381-2613-5

Zugl. / Approved by: München, TU, Diss., 2010

Preface

The PhD dissertation presented here is the fruit of the work done from August 2007 until August 2010 at the Institute of Hydrochemistry of the Technische Universität München.

My warmest thanks are reserved to Prof. Reinhard Niessner who gave me the opportunity to work in his aerosol group. After my master thesis, which I did at his institute as well, he invited me to continue as PhD student under his supervision. I want to thank him for the trust and for the possibility to work on my own project dealing with such an interesting and actual subject. His door was always open for questions and suggestions, and I have tried to learn as much as possible from his deep knowledge of aerosol science and analytical chemistry.

Our aerosol group was a good environment to work. My thanks go to Markus Knauer who collected all the aerosol samples that I used for the biofuel project, and to Carsten Kykal who introduced me to SMPS and ELPI measurements. My colleagues Johannes Schmid, Gabriele Hörnig and Benedikt Grob provided an enjoyable atmosphere and useful discussions.

During my PhD work I was tutor of one master thesis, two bachelor thesis and several practicals. My thanks go especially to Dominik Deyerling for his support during the pyrene experiments and to Jan-Christoph Wolf for the contribution during the benzo(a)pyrene series as well as for the synthesis of standards for the analysis of 1- and 3-nitrobenzo(a)pyrene. Their precious work and their useful collaboration made possible a part of the results presented in this dissertation.

I also wish to acknowledge Dr. Diether Rothe and Frank Ingo Zuther who made the sampling at MAN possible and Dr. Jürgen Blasnegger who was the coordinator of the biofuel project.

I am very grateful to Susanne Mahler for the HPLC measurements but also for the moral support and to Sebastian Wiesemann who furnished excellent technical support in every field, from O-rings to motorbikes.

The list of people that I have met in the last three years at the IWC is quite long. Among others I want to acknowledge Dr. Christoph Haisch and Dr. Natalia Ivleva and, in open order, my former colleagues Philipp Stolper, Caroline Peskoller, Christian Cervino, Katrin Kloth and Anne Wolter for the warm welcome in my first months when I was a newcomer. Now I should be the PhD student with the highest length of service, and I was pleased to work with Clemens Helmbrecht, Gerhard Pappert,

Simon Donhauser, Jimena Sauceda, Susanne Huckele, Christina Mayr, Xaver Karsunke, Maria Knauer, Martin Rieger Susanna Vasac, Klaus Wutz and many others not explicitly listed here.

I also want to thank my parents Paolo and Lucia, my brother Francesco, and every member of my family who on different occasions and in different ways contributed to the final outcome of the present dissertation. Finally, I have especially to thank my girlfriend Mascha. She helped me a lot during the difficult periods of my PhD work and she was a constant source of motivation and support.

Relevant publications

Carrara M.; Wolf J.C.; Niessner R. "Nitro-PAH formation studied by interacting artificially PAH-coated soot with NO_2 in the temperature range 295–523 K". *Atmospheric Environment* **2010,** 44, 3878-3885

Carrara M.; Niessner R. "Impact of a NO_2-regenerated diesel particulate filter on PAH and NPAH emissions from an EURO IV heavy duty engine". *Journal of environmental monitoring,* Submitted

Carrara M.; Knauer M.; Niessner R. "Emission of PAH, nitro-PAH and carbonyls by operating four modern vehicles with fossil fuels, biofuels and their blends". *Journal of environmental monitoring,* Submitted

Abstract

The present dissertation investigated the interaction of soot-adsorbed PAH with NO_2 both in laboratory model systems and in real diesel engines exhaust. Scope of the experiments was to explain to which extent modern diesel particulate filters (DPFs) could support the formation of highly toxic and mutagenic nitrated polycyclic aromatic hydrocarbons (NPAHs) by nitration of pre-existing PAHs. In addition, the use of biofuels in different vehicles was investigated with respect to the effects on NPAH emission.

This work focused on the heterogeneous reactions which occur on soot particles by exposing laboratory produced pyrene (PYR)- or benzo(a)pyrene (BaP)- coated spark discharge soot particles to varying concentrations of NO_2 (0.1 ppm - 120 ppm) and temperatures (20°C – 300°C) while following the formation of products over time. The sole reaction product observed by nitration of PYR-coated soot was 1-nitropyrene (1-NPYR), which increased linearly with reaction time. In comparison, BaP was more reactive than PYR and its main nitration product was 6-nitrobenzo(a)pyrene (6-NBaP) with trace amounts of 1-NBaP and 3-NBaP. Highest 1-NPYR concentrations on particles were detected at 100°C, and at higher temperatures particulate 1-NPYR decreases considerably. A similar trend was observed in a DPF-simulation system with BaP-coated soot. In this case highest 6-NBaP concentration on particles was detected at 150°C. Backed by corroborating results from separate gas/solid phase partition experiments, it is likely that the newly formed 1-NPYR and 6-NBaP became transferred from particle to gas phase at higher temperatures. Results from this work confirm the production of 1-NPYR and 6-NBaP in particulate and gas-phase under conditions encountered in DPFs. Experiments with a real diesel engine, showed that in real DPFs the PAH and NPAH degradation competes with the NPAH formation. Therefore, though the total amount of PAH and NPAH decreased with the use of the DPF, some compounds became degraded while others were newly formed. At low load, after the DPF 1-NBaP or 6-NBaP increased approx. by a factor 15 and the super-mutagen 1,6-dinitropyrene was completely newly formed.

In regard to emissions using biofuels, in an EURO V truck and in an EURO V diesel car, NPAH concentrations in the exhaust were not significantly different from those of fossil fuel. In an EURO V otto-cycle car and an the EU III A tractor, emissions from biofuels were sensibly lower.

Index

1 Introduction

For decades air quality has been a major issue for scientists, which have been required to furnish knowledge and solutions to improve it. Air quality is a measure of the concentrations of gaseous pollutants and size or number of particulate matter (1). One of the most exciting and interesting aspects of air quality is the study of aerosol particles. Atmospheric aerosol particles are generally defined as a suspension of liquid or solid particles in a gas, with particle diameters in the range of 1 nm – 100 µm (lower limit: molecules and molecular clusters; upper limit: rapid sedimentation) (2). They produce effects on the troposphere, climate and human health (3). The most easily understood interactions between aerosols and climate are the scattering and the absorption of solar and terrestrial radiation (4). They also impact the hydrogeological cycle by acting as condensation and ice nuclei for clouds thus effecting precipitation (2,4). Aerosol particles are also involved in a great amount of heterogeneous and multiphase reactions which affect the gaseous composition of the atmosphere (5).

Atmospheric aerosol particles are generated from a wide variety of natural and anthropogenic sources. The principal natural sources are volcanic eruptions, stone erosion and dust re-suspension, wildfires, marine aerosols and biogenic emissions. The principal anthropogenic sources are combustion processes and industrial activities (6). Among anthropogenic sources, emissions from traffic and diesel engines in particular, are of great concern for air quality and represent a major part of the total emitted particulate matter (PM) (7,8). To characterize the aerosol as a function of size, the concept of the aerodynamic diameter has been introduced. This is defined as the equivalent diameter of a spherical particle (density of 1 g/cm^{-3}) that has the same settling speed of the particle in matter. The particles with diameters below 10 µm (PM10) represent the thoracic and/or breathable fraction and are harmful to humans (9). PM10 is involved in the adsorption processes of other compounds, in gaseous pollutant transformations and in long-range transport phenomena. For these reasons it has been chosen as a parameter to estimate air quality concerning particulate matter. In Europe the highest concentration allowed is 50 $\mu g/m^3$ (24 hours average) and can be exceeded only 35 times per year (10). Up until now PM10 is the only indicator with legal relevance concerning aerosols in the atmosphere.

Diesel engines emit particles in the range of 20 nm - 400 nm (11) which are classified as fine and ultrafine particles. They are numerous but, due to their small size, they have extremely low mass (12,13). This renders them practically invisible when particle concentration is measured using the PM10 mass concept. For this reason, even if diesel emissions contribute substantially to atmospheric particle concentrations in urban areas, has been concluded that PM10 mass is a poor indicator of exposure to diesel particles (8,14). Therefore the new emission limits for diesel vehicles which will be introduced in Europe by 2015, also consider the emitted particles number concentration (15).

Diesel PM is widely considered as a probable human carcinogen (16,17). It represents a complex mixture of organic and inorganic materials where the inorganic part mainly consists of fine soot particles produced during the high temperature combustion of fuel (18). Much is still unknown about the toxicology of ambient particulates and the effects that can be directly attributed to diesel particulate matter are still uncertain. However, it is sure that exposure to particles <PM1 (particles with an aerodynamic diameter lower than 1 µm) and to ultrafine particles causes an increase in risk for respiratory and cardiovascular diseases (19-21). Ultrafine particles such as diesel PM, have been found to be more toxic per unit mass than coarse particles (22). Several classes of organic components such as polycyclic aromatic hydrocarbons (PAH), nitrated PAH (NPAH) and aliphatic hydrocarbons can adsorb on the soot surface thus enhancing its toxicity (23-25). Both the organic and the PM components cause oxidant lung injury. It is known that the diesel PM induces alveolar epithelial damage, alter thiol levels in alveolar macrophages and lymphocytes. The first can be activated in the production of reactive oxygen species (ROS) and pro-inflammatory cytokines which can then cause DNA adducts (25,26). The organic component, on the other hand, is shown to generate intracellular ROS, leading to a variety of cellular responses including cellular death (apoptosis) (26).

Diesel particulate filters (DPF) have evolved into a promising technology that reduces harmful diesel emissions. Most commonly, these filters consist of ceramic monoliths with alternating flow channels that are closed at the ends in order to force the exhaust flow to pass through the porous wall of the honeycomb filtration medium (27,28). In these so called "wall flow" filters the solid particles are deposited in the pores with a filtering efficiency typically between 90 % - 99%. Consequence of this filtration method is that the pores become plugged with soot leading to increased backpressure that compromises the proper engine operation. Other possible DPF designs have a relatively open structure. In such filters particulate matter is found throughout the filter structure and not just on one side of the filter surface, as with wall-flow filters. This mechanism, called deep-bed filtration (29), gives less problems in terms of increasing backpressure but compromises filtering efficiency which is typically between 30% to 50%. To remove the soot particles trapped in the filters, it is necessary to regenerate them periodically at elevated exhaust temperatures, by, e.g., first passing the exhaust gas through a diesel oxidation catalyst designed to convert NO to NO_2, which is then used to assist the filter regeneration by enhancing soot oxidation (27).

PAHs are often already present in the fuel and they may survive the combustion process while others undergo de-novo formation (30), therefore considerable amounts of PAHs may reside on the soot particles. For this reason, the filter structure may act as a reaction chamber for the nitration chemistry that would result in the post-combustion formation of NPAHs. This would be a highly unwanted effect because NPAHs, even in concentration 10 – 100 times lower than those of PAHs, are much more toxic and mutagenic than their parent material (31-34). The wide range of temperatures at which different DPFs operate, could have an effect either on the reaction kinetic, or on the particle / gas-phase partitioning of NPAHs. On the other hand, also a further oxidation of engine-out PAH and NPAH may be caused by the high NO_2 concentration and high temperature (35). It is then necessary to demonstrate to show which of these competing processes prevail.

Although heterogeneous gas-particle degradation of PAHs has already been studied (36,37) and real DPFs have been tested with respect to NPAH formation (35,38), information on the nitration kinetics with focus on reaction products is still lacking. Moreover, measurements done by the author of this work analyzing the NPAH emission from different vehicles and different fossil and bio-fuels, have drawn attention to a high 1-nitropyrene emission from vehicles equipped with a DPF. Unfortunately, those results were affected by the absence of a systematic investigation of the NPAH artifact formation during sampling. This is problematic as NPAH artifacts can be an issue in diesel exhaust sampling because collected PAHs can be nitrated on the sampling filter by the NO_2 emitted from the motor and/or from the exhaust aftertreatment system (39,40). This has already been observed in engines equipped without a DPF where the engine-out NO_2 concentrations are sufficiently high to support this chemistry. In the case of NO_2-regenerated DPFs the NO_2 concentrations are even higher. Pyrene (PYR)$\rightarrow$ 1-nitropyrene (1-NPYR) conversions of up to 30% and 45% have been determined without and with DPF, respectively (35).

The core of the present work is the interaction of laboratory generated PAH-coated soot particles with NO_2 and the resulting NPAH formation. This is investigated under varying NO_2 concentration and temperatures with a focus on the reaction products, especially NPAHs. A part of the experiments was designed to investigate the artifact formation of NPAH under sampling conditions. Others were designed to understand whether or not a post-combustion formation of NPAH can take place within a DPF structure. The aerosol particles used as model soot carriers for PAH can be regarded as a reasonable proxy for the core of real diesel soot (41). In addition to the laboratory experiments which provide the basic knowledge to understand the nitration reaction, work done exploring the effects of DPFs in real diesel engines gives a more complete description of the topic.

A second topic treated in this study was the unregulated emission by the use of biofuels in different vehicles. In fact, there is a finite amount of fossil oil which can be applied as fuel. Also an enhancement in the greenhouse effect, hence earth atmosphere heating, implies that alternative energy sources are needed. By use of biofuels and biofuel blends, it is possible to partially substitute fossil fuel. Though, up to now, besides the still open questions on the use of arable land and edible crops to produce biofuels, there is just little knowledge about to which extent the addition of biofuel to fossil fuel will change the emission of compounds such PAH and NPAH.

To get more detailed information on the emission rates of the different engine exhaust components, samples were taken from dilution tunnels at vehicle roller test benches in Graz and Vienna during defined test cycles. A heavy duty vehicle, a tractor motor as well as passenger cars were examined during this study. To determine differences in the emission rates, fossil diesel, biodiesel and vegetable oil as well as mixtures with different amounts of biofuels in fossil fuel were investigated. The results presented here are only a part of a bigger study (called from here on "biofuels project") which investigated both regulated and unregulated emissions. The author of this work was responsible of the NPAH analysis: therefore only results concerning NPAH will be presented.

2 Theoretical background

2.1 Diesel exhaust emissions

The great majority of vehicles used around the world rely on 4-stroke internal combustion engines. At the core of a combustion engine is the piston within a cylinder and two classes of valves (intake and exhaust). Spark-ignition engines, also known as otto-cycle engines, contain a spark plug to trigger the combustion reaction of the gasoline air mix. Diesel engines do not have a spark plug, instead they rely on auto ignition of the fuel (42). In a diesel engine, the fuel is injected directly into either the cylinder and the fuel burns as a diffusion flame attached to the fuel injector. As diesel fuel aerosol has good self-ignition properties, the fuel droplets ignite spontaneously (43). The fuel combustion increases the gas temperature, raising the pressure in the cylinder, which causes the piston to be driven down generating power to propel the vehicle. A fraction of a second before the piston reaches the bottom of its travel, the exhaust valve opens and the exhaust gases are pushed out by the rising piston through the exhaust valve, through an eventually included after-treatment system, and through exhaust pipe into the atmosphere (42).

2.1.1 Engine combustion

During combustion, fuel composed of carbon and hydrogen is oxidized by oxygen (O_2) to form carbon dioxide (CO_2) and water (H_2O) by following the reaction (44):

$$C_nH_m + \left(n + \frac{m}{4}\right)O_2 \rightarrow nCO_2 + \frac{m}{4}H_2O$$

Normally the percentage mass composition of diesel fuel is 86.4% of carbon and 13.1% of hydrogen. O_2 necessary for the combustion is fed as air which at 22°C and 50% relative humidity has a O_2 mass percentage composition of 23%. Therefore, under these circumstances, for the complete combustion of 1 kg of diesel fuel 14.6 Kg of air are necessary (15,44).

The parameter λ (lambda) gives the ratio between the mass of air at disposal for the combustion and the stoichiometric need of the reaction. In diesel engines the combustion is not ideal and they are operated over-stoichiometric with λ that varies between 2 and 4 at medium load and between 1.3 and 1.8 at full load (15). At very low load, a λ value of 11 can also occur (45). Modern engines are electronically controlled to optimize the amount of fuel and air necessary for the best combustion depending on the request of power.

2.1.2 Gaseous emissions

Because the combustion is not ideal, compounds others than CO_2 and H_2O are produced in a mass percentage which can reach even 1% (46). Therefore the exhaust emission represents a complex mixture where the main gaseous products of diesel exhaust are N_2 CO_2, O_2, and water vapor, but also pollutants such as carbon monoxide (CO), hydrocarbons (HC) and nitrogen oxides (NO_x) are present. Depending on the sulfur content in the fuel and in the lubrification oil, within the combustion chamber sulfur dioxide (SO_2) can be also formed. SO_2 can react further to form sulfuric or to sulfate. In western countries, the actual legislation reduced the maximum tolerated sulfur content in fuels drastically to 10 ppm but average sulfur content is typically 3 ppm – 5 ppm (47); therefore the emission of SO_x from diesel vehicles in western countries are nowadays considered a minor problem.

One of the major pollutants produced in combustion engines are NO_x which represent the sum of nitrogen monoxide (NO) and nitrogen dioxide (NO_2). Actually, since fuel oxidation occurs in the presence of N_2 (ambient air is composed of 78% N_2), the endothermic reaction of NO formation takes place. In diesel engines the fluid motion and its time-scale are such that the formed NO_2 efficiently reduces back to NO, which turns out to be the only contributor to engine-out NO_x (48). A fraction of the in-cylinder NO will oxidize to NO_2 along the exhaust pipes depending on temperature and O_2 residual concentration, but the total NO_x emission will be the same. The equilibrium between NO and NO_2 at temperatures above 500°C is completely shifted to the NO side. At lower exhaust temperatures of about 250°C – 350°C NO_2 usually represents 2% - 15% of the total NO_x tailpipe emissions (49). Regarding the influence of exhaust O_2 concentration, lowering the load increases the NO_2 to NO_x ratio because of the more lean conditions. In fact, at low load, λ values are higher than at high load, thus increasing the exhaust O_2 which can oxidize NO to NO_2 to a maximum 30% of the total NO_x emissions (45).

In the past, experiments have revealed most of the phenomena related to combustion in internal combustion engines (50). Nonetheless, to date, no experimental technique has been able to clarify all the relevant chemical reactions. For a more detailed understanding of the chemical processes, investigators have turned to computer models (48). It would be beside the purpose of this work to review all these models, but some key information can be given specifically regarding the formation of NO_x, one of the most relevant pollutants deriving from diesel engines, which also play a key role in the soot emission management.

Current NO_x models for diesel engine simulation, assume the formation of NO_x via the extended Zeldovich mechanism. It comprises three reactions (48,51):

$$N_2 + O \leftrightarrow NO + N$$

$$N + O_2 \leftrightarrow NO + O$$

$$N + OH \leftrightarrow NO + H$$

This mechanism is often referred to as the thermal mechanism, because of its great temperature sensitivity. In practice, it occurs above 1400°C when the forward reactions lead to a net oxidation of N_2 to NO (48). The other two formation pathways have been proposed for fuel-rich mixtures and for lean conditions, respectively. In fuel-rich mixtures the NO formation is initiated by the reaction of N_2 and •CH radical, which is abundant, to form HCN that reacts further to NO (44,48,52). This mechanism may play a role in high speed engines, where the highly swirling air penetrates the fuel spray and transports nitrogen into fuel-rich zones. On the contrary, this mechanism should be irrelevant in low speed engines, where the fuel spray penetrates into the mass of air, and nitrogen remains on the fuel-lean side. In fuel-lean mixtures, NO may form via the mechanism known as N_2O-intermediate mechanism; these reaction are

$$N_2 + O + M \leftrightarrow N_2O + M$$

$$N_2O + O \leftrightarrow NO + NO$$

where M is a third body (51). This mechanism has been indicated as an additional source of NO in internal combustion engines during the early expansion stroke, being promoted at high pressure (48).

Experiments to evaluate the impact of a driving cycle on NO_x emission of different vehicles have shown that heavier cars produced more NO_x than the lighter ones. That is because the engines had to run at higher loads to satisfy the power needs of each cycle. As a higher load induces a higher combustion temperature and NO_x emissions increase with temperature, NO_x emissions are higher at higher loads (53).

As it is well known, the reaction between NO_x and some hydrocarbons in the presence of sunlight produces photochemical smog with the consequent formation of ozone. Diesel engines, among other sources like otto-cycle engines or industrial combustion processes, are responsible of the emission of both these pollutants. HC are products of incomplete combustion and the variety of compounds that can be emitted is huge (54,55). That is also due to reactions that can further transform HC. Cracking (division of a long carbon molecule into two shorter molecules), polymerization (combination of 2 olefinic molecules into a bigger olefinic molecule) and reformation (catalytic conversion of paraffin or naphthalenes into molecules of aromatic structure by means of pyrolysis) are likely to occur at certain pressures and temperatures in the engine (56).

2.1.3 Soot emission

The diesel engine combustion process produces highly carbonaceous material known as soot, which is one of the major air pollutants (57). Soot is technically defined as the black solid product of incomplete combustion or pyrolysis of fossil fuels and other organic materials. In a diesel engine, fuel is sprayed into the cylinder at high pressure. The sprayed diesel fuel droplets do not mix completely with the abundant oxygen at a molecular level, which results in incomplete combustion. Actually, oxygen transport occurs by diffusion through the flame front, and this type of reaction zone is therefore called a "diffusion flame" (43). Soot is then formed from unburned fuel, which nucleates from the vapor phase to a solid phase in fuel-rich regions at elevated temperatures. Hydrocarbons or other available molecules may condense on, or be absorbed by soot depending on the surrounding conditions (58).

A clean combustion with low soot emissions requires possibly high temperatures. Despite of that, PM emissions increase with engine speed where temperatures are generally higher than at low engine speed. This is not a contradiction because as the load demand increases, the combustion requires a higher fuel/air equivalence ratio and consequently PM emissions are higher (53).

2.1.4 Mechanism of soot formation

The formation of soot, i.e., the conversion of hydrocarbon fuel molecules containing few carbon atoms into a carbonaceous agglomerate containing some millions of carbon atoms, is an extremely complicated process. It is a kind of gaseous solid phase transition where the solid phase exhibits no unique chemical and physical structure (57).

The evolution of liquid or vapor phase hydrocarbons to solid soot particles and products involves four commonly identified processes: pyrolysis, nucleation, surface growth and agglomeration (58,59). Finally, it is not excluded that newly-formed soot is oxidized back to gas-phase.

By now, two main hypothesis explaining the soot formation in combustion processes are available in the literature: the so cold "ethyne" or "acetylene" hypothesis and the "aromatic" hypothesis (60). However both theories concern complex processes encompassing formation of precursors such as PAH, through radical reactions of small molecules and/or fragments. According to the "ethyne" theory, fuel molecules are pyrolysed into small unsaturated molecules where the main constituent is ethyne (29). That is mainly due to ethyne`s thermodynamic stability which increases with increasing temperature whereas the stability of all other types of hydrocarbons decreases (60). The ethyne polymerises to polyethyne and, eventually, polycyclic structures are formed by ring closure. These polycyclic structures are called platelets and are, in fact, small graphite-like sheets that could be seen as the primary building. The platelets stack together to form crystallites, which subsequently stack together themselves to form turbostratic particles (29), where the platelets and crystallites are randomly orientated around the center of the particle.

The size of the particles increase due to particle coagulation and surface growth is instigated by the addition of precursor gas-phase molecules. The surface growth fills the space between the coagulated particles, and because of this, the coagulated particles remain spherical (29). The last step of diesel soot formation is then the aggregation to bigger conglomerates of irregular form, as it is shown in Figure 2-1.

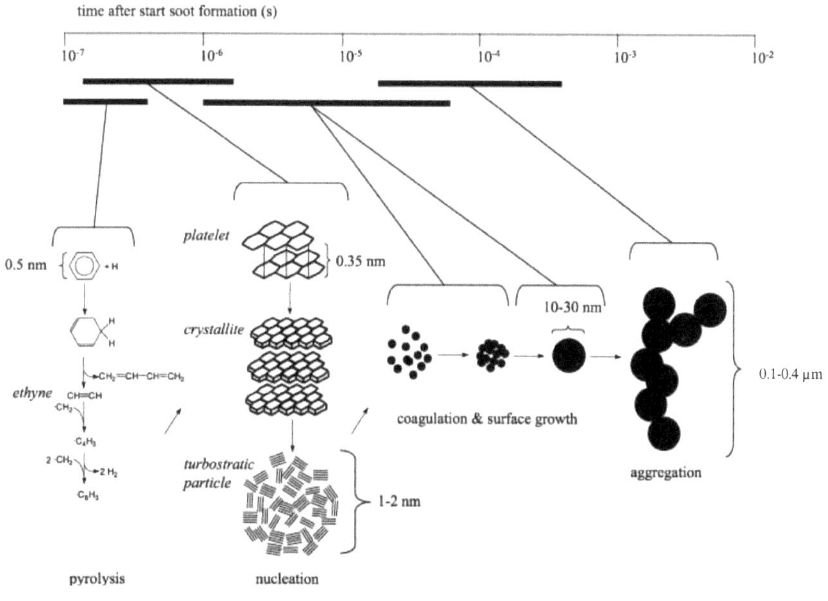

Figure 2-1. Mechanism of the formation of Diesel soot particle (29).

2.1.5 Soot structure and reactivity

It is well know that diesel soot particles are usually chain-like agglomerates composed of several hundreds of primary particles (61). The cohesion is due to surface forces to which adsorbed material plays a part (62). A transmission electron microscopy (TEM) photo of diesel soot particles is provided in Figure 2-2. Primary particles are spherical and have typical diameters in the order of 10 - 30 nm comprising crystalline and amorphous domains (63). The graphite-like crystalline domains typically consist of 3 - 4 graphene layers, with average lateral extensions of up to 3 nm and interlayer distances of about 3.5 Å and can be regarded as highly disordered graphitic lattices.

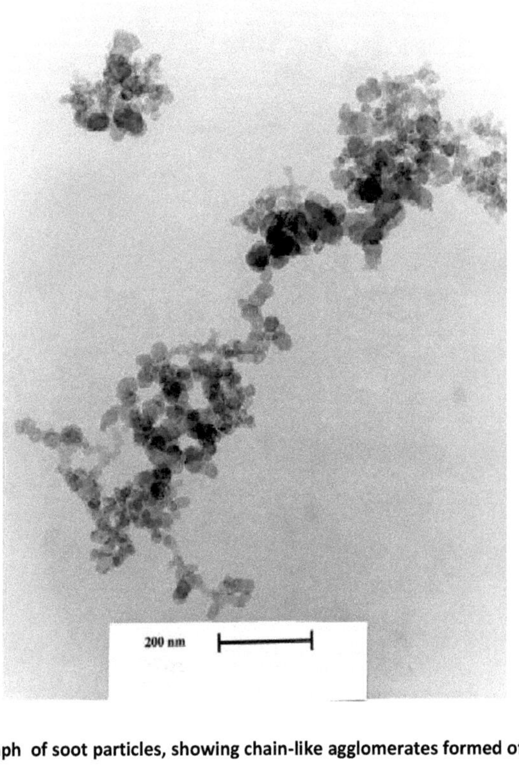

Figure 2-2. Micrograph of soot particles, showing chain-like agglomerates formed of primary spherical particles (62).

The amorphous domains are composed of polycyclic aromatic compounds, which can be regarded as graphene layer precursors in irregular or onion-like arrangements (fullerenoid structures) (63), and other organic and inorganic components (aliphatics, metal oxides, etc.).

It is also possible to find a mixture of sp^2 and sp^3 carbon with no long-range order (64). Analysis of TEM pictures showed that surface growth platelets do not orientate themselves in a turbostratic manner, instead they are orientated perpendicular to the radius of the particle. This orientation gives to the soot particles a better ordered outer shell (25), which is thermodynamically more stable than the turbostratic inner core (61). The structure of the diesel soot agglomerates can be described by the concept of fractals. Based on the relation between mobility diameter and mass, typical values of the fractal dimension are about 2.3 (65).

Because these structures are by far not spherical, physical properties such as density are strongly influenced. The effective density of the particles in terms of the electrical mobility diameter has be estimated to be 0.8 g cm^{-3} for a diameter of 100 nm and decreases with increasing particle size reaching about 0.3 g cm^{-3} at 300 nm. This means that at an electrical mobility diameter of 300 nm, the aerodynamic diameter is only about 100 nm (65). (For a more detailed description of the electrical mobility diameter and aerodynamic diameter, see section 3.2.1). Anyway, after many years of research, there have been large deviations in the reported property values for diesel soot, especially regarding porosity and permeability and therefore apparent density which range from 0.14 g cm^{-3} to 1.0 g cm^{-3} (66). This disagreement is understandable, taking into account that the properties of the diesel soot depend on the nature of the combustion, engine type and engine operating point, which can significantly influence the structure and therefore the density.

The actual physical and chemical structure of soot, its elemental composition, and the ratio of crystalline graphite-like to amorphous organic carbon depend on the starting materials and conditions of the combustion or pyrolysis process: fuel type, fuel/oxygen ratio, flame temperature or combustion temperature, residence time etc.. For instance TEM measurements have revealed an increasing diameter of primary soot particles from 12.9 nm to 21.7 nm for a heavy-duty diesel engine due to increased engine load from 25% to 75%, or a reduction of the same order of magnitude by increasing the injection pressure (67). Moreover due to the implementation of the combustion process, modern engines seem to produce soot with smaller primary particles than older ones. In old diesel engines the precursor molecules grow from 1 nm – 2 nm to 10 nm – 30 nm before aggregation or chain-forming coagulation takes place. This is only possible if precursor molecules are readily available (rendering a rapid growth), or the surface of primary particles is less functionalised and therefore less prone to aggregation. The small particle size and the rough surface of Euro IV soot suggest that in the modern engines, due to the improved combustion of the diesel fuel, fewer precursor molecules are produced leading to slower growth of the primary particles (68). Additionally, the surface may be strongly functionalised with surface functional groups (SFG) so that the particles coagulate before they grow in size.

Regarding the reactivity of diesel soot, a graphitic soot primarily exposing graphene basal planes is generally less susceptible to oxidative attack (69). Actually, since a graphitic carbon is characterized by layer planes with larger dimensions of the edges parallel to the basal plane, by virtue of geometry, larger layer planes contain fewer carbon atoms at edge sites relative to basal plane sites (70). Carbon atoms within basal planes, surrounded by other carbon atoms, exhibit a far lower reactivity towards oxidation than those located on the edges. In fact, carbons atoms at the edge sites can form bonds with chemisorbed oxygen due to the availability of unpaired sp^2 electrons, while carbon atoms in basal planes are more aromatic having only shared π electrons to form chemical bonds (71).

2.1.6 Diesel particulate composition and size distribution

Diesel particles are a complex mixture of elemental carbon (EC), organic carbon (OC), sulfur compounds, and other species (65). The composition of diesel particulate matter depends on where and how they are collected (29). At the cylinder outlet and in the tailpipe, where the temperatures are high (>350°C), most of the volatile materials (HC, sulfuric acid or volatile PAH) are in the gas phase. As soon as the exhaust gas leaves the vehicle, it is diluted by ambient air and cools down. Nucleation, condensation, and adsorption transform the volatile materials and the aerosol particles, changing the composition and size distribution of the emitted PM.

Moreover, surface oxygen groups are inherently produced as intermediates upon partial oxidation. Oxygen functional groups include phenolic, carbonyl and carboxylic acids groups (69). The most recent literature available on the analysis of diesel particulate, is provided in Figure 2-3. A tiny fraction of the fuel and atomized lubrification oil escape oxidation and appear as volatile or soluble organic compounds (generally described as the soluble organic fraction, SOF) in the exhaust (72). Metal compounds in the fuel and lubrification oil lead to a small amount of inorganic ash. Other authors characterized and analyzed the carbonaceous materials and found that EC accounted of ca. 30% in mass, OC about ca. 35% while about 30% remained unidentified (73). Of the extractable and elutable material most, approx. 85%, was unresolved by gas chromatography (GC). About 70% of the resolved mixture cannot be identified; thus, the roughly 100 identified compounds constitute less than 5% of the OC in the PM (73,74).

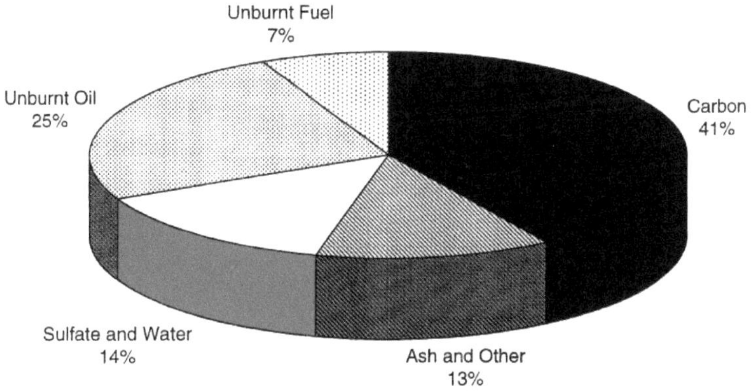

Figure 2-3. Typical diesel PM composition(heavy-duty diesel engine, transient cycle) (72).

In Figure 2-4 the size distributions of diesel exhaust particles (number and mass weighting) are shown. The distributions are tri-modal. Lognormal distributions usually fit actual data very well. The concentration of particles in any size range is proportional to the area under the corresponding curve. Most of the particle mass exists in the so-called accumulation mode in the 100 nm - 300 nm aerodynamic diameter range. This is where the carbonaceous agglomerates and associated adsorbed materials reside. The nuclei mode typically consists of particles in the

3 nm - 30 nm aerodynamic diameter range. This mode usually consists mostly of volatile organic and sulfur compounds that form during exhaust dilution and cooling (72). These small particles may also contain solid carbon, ashes and metal compounds that can serve as condensation nuclei also for fuel droplets. The nuclei mode typically represents 1% - 20% of the particle mass and more than 90% of the particle number. The coarse mode contains 5% - 20% of the particle mass. It consists of smaller particles (accumulation mode) that have been deposited on the cylinder and exhaust system surfaces forming bigger aggregates which can later be re-emitted.

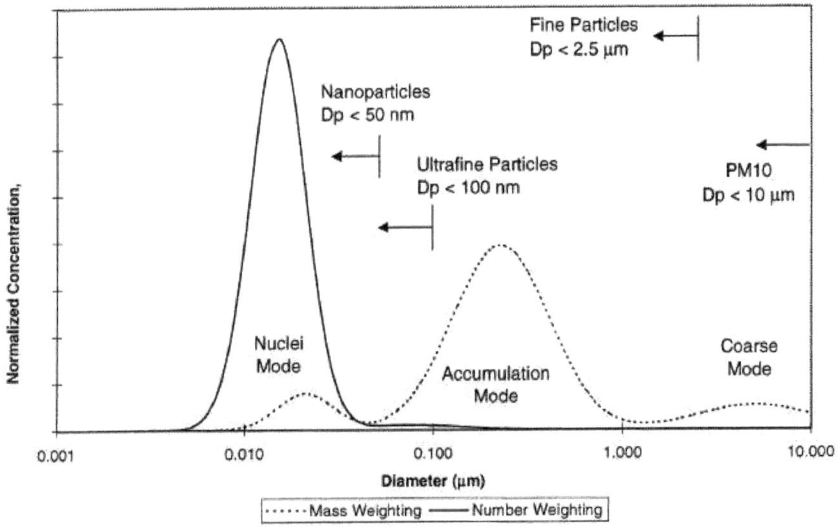

Figure 2-4. Typical engine exhaust size distribution. Both mass and number weightings are shown (72).

2.2 Diesel exhaust emission control:

The lower CO_2 emissions and fuel consumption of diesel engines favor them in comparison to otto-cycle engines. Beside the use in heavy-duty vehicles, their utilization in passenger cars is increasing. Despite of that, the long-term effects of diesel engines in transportation and energy generation systems are still under debate. Critical factors that will influence the fate of diesel engines are the soot and the NO_x emissions. Therefore, systems to control the soot and NO_x emissions due to their environmental and health concerns are needed. This requires the development of efficient diesel engines and suitable aftertreatment systems.

2.2.1 Engine optimization

The pollutant emission reduction can begin directly at the origin, in the fuel refining and in the engine combustion. As already stated, sulfur-free fuels can dramatically reduce the emissions of SO_2 and sulfates. Moreover, automakers are requesting diesel fuel with high cetane numbers to improve cold start feasibility and reduce emissions (e.g. white smoke, the mixture of unburned or partially burned fuel and water vapor) (42).

Optimization of in-cylinder fuel combustion can reduce PM and/or NO_x emissions. However, many engine-side technologies developed so far, tend to decrease PM emission while increasing NO_x emission, and vice versa. For example, retarding the fuel-injection timing can be effective to reduce NO formation. However, this usually results in an increase of soot production (75). A further example is the increased fuel injection pressure, which although it can decrease soot emissions, it can also cause higher NO_x emissions at the same time (76). That is the so called "trade-off" effect that puts the engine optimization for the reduction of soot with that of NO_x into competition.

Exhaust gas recirculation (EGR), is an emission control system that regulates the amount of engine-out exhaust gas that is mixed into the intake stream of the cylinder (77). It has a positive effect on NO_x emissions because it lowers the temperature and the oxygen content of the air used for the combustion. On the other hand, for the same reasons, EGR leads to a less complete fuel combustion producing higher amounts of particulate.

In real engine emissions, the ultimate goal of reducing pollutants under the most recent emission limits could not be accomplished by the engine optimization alone. Improvements of the engine-side are just a part of the control strategy. In order to achieve the emission limits, several after-treatment systems can be used without penalties of engine performances by operating the engine within optimum conditions. The next sections will provide an overview on aftertreatment diesel emissions control strategies.

2.2.2 After-treatment systems for NO_x emission control

Selective catalyst reduction (SCR) is the leading NO_x control strategy for both light-duty and heavy-duty applications. SCR works by reducing NO_x using an ammonia reductant (78). As ammonia is a toxic gas, the actual reducing agent used in motor applications is urea. Urea is commercially available and is both ground water-compatible and chemically stable under ambient conditions (28). The ammonia is then in-situ created by pyrolysis. It reacts selectively with NO_x on a catalyst as reducing agent under the lean and low temperature (250°C – 500° C) condition by one of the following chemical reaction paths (75):

$$NO + NH_3 + 1/4\ O_2 \leftrightarrow N_2 + 3/2\ H_20\ (slow)$$

$$1/2\ NO + NH_3 + 1/2\ NO_2 \leftrightarrow N_2 + 3/2\ H_20\ (fast)$$

If the fast chemical reaction path could be applied in the SCR system, the overall conversion efficiency improves, especially at low temperature (≤ 300° C).

Regarding the catalyst itself, in Europe SCR catalysts are based on vanadium which is cheap and sulfur tolerant. If DPFs are used with SCR systems, such as in Japan and it the USA, copper- or iron-zeolites are preferred due to their better high temperature durability needed during DPF regeneration with fuel post-injection which can expose the SCR catalysts to temperatures up to 800°C (78).

Urea SCR's main disadvantages are related to the requirement of onboard storage of the urea reductant and the need to heat the urea during cold weather. The issue of enforcing re-supply of the urea reductant by drivers must also be addressed (79).

Another strategy for NO_x emission control can be attractive because it is relatively cost saving and it does not have the complication of urea storage and injection. It is the case of the NO_x storage catalysts working at lean engine operation also known as lean NO_x traps (LNTs). LNTs switch between two modes of operation. Most of the cycle is spent trapping NO_x during lean conditions as a nitrate on alkali earth materials (such as barium oxide (BaO)). Periodically, control systems cause the engine to run fuel rich, and unburned fuel in the exhaust is used to liberate the stored NO_x and reduce it to N_2 on an integrated catalyst (rhodium or platinum) (28,75,78,79).

2.2.3 After-treatment systems for diesel particulate emission control

Particulate emission control covers a wide range of measures to reduce both the solid (soot) and the liquid particulate (droplets) emissions. Liquid particulates can easily be reduced by regulating their precursors e.g. by reducing the fuel sulfur level and installing diesel oxidation catalysts to oxidize the HC and fuel micro-droplet emissions which are the main constituent of liquid particulate due to condensation (80).

Regarding soot emission, DPFs have become possibly the most important and complex diesel emission control device (80). Car manufactures could achieve the EURO IV emission limits (valid until 2009) without using DPFs and just with the improvement of the combustion systems. As diesel emission regulations become more stringent (EURO V or EURO VI), DPFs represent the only solution to fulfill the requirements for diesel particulate emissions.

DPFs are an important and highly complex type of multifunctional chemical reactor where, over a wide range of temporal and spatial scales, filtration, chemical reactions and material transformations occur. Because of the central importance of DPFs in diesel exhaust aftertreatment, in the last ten years the research and development of DPFs has been tremendous and produced a great variety of systems with different structures, materials, filtration and regeneration mechanisms. The next section summarizes the most commonly used systems and delineates the most important trends.

2.2.4 Diesel particulate filters

The first car mounting a DPF that appeared on the market, was a Peugeot in 1999 (78). The system created was based on three main groups of elements. 1) A specific software dedicated for regeneration control and system diagnosis. 2) A filtration medium made of porous silicon carbide (SiC) located downstream of an oxidation catalyst equipped with differential pressure sensor and two temperature sensors. 3) A fuel additive in a specific tank to produce a cerium-based fuel born catalyst (FBC) (81). To start or stop the regeneration process the mass of soot in the filter was continuously monitored by measuring the filter pressure drop. The regeneration process began with an in-cylinder gas temperature increase by specific management of multiple injection and air boost pressure. The amount of HC in the exhaust gas was increased by multiple injection tuning. The oxidation catalyst in front of the filter assured the catalytic post combustion of the HC. These two actions raised the filter temperatures up to 300°C (81) and, aided by the effect of the FBC (cerium based), regenerated the filter completely.

Filter regeneration is a major issue. With the accumulation of PM in the filter medium, the pressure drop across the filter increases which in-turn exerts a back pressure on the engine exhaust, resulting in gradually poorer performance of the diesel engine in terms of higher CO emissions, poor fuel efficiency, and increased PM production (82). Therefore it is necessary to clean the filters, which is usually done by thermal means.

Soot starts to burn spontaneously at a temperature of about 600°C (83). However, such temperatures are not very often available in the exhaust pipe; therefore, it is necessary to use technical means to control the regeneration of the filter. Moreover, a trap should be able to regenerate during all driving conditions without the intervention of the driver (84) following the principle *"fit & forget"* .

Besides the already explained DPF from Peugeot, many others approaches have been used in PM filtration. Filter systems can be characterized and sorted considering one of these features:

1) Design of filter & mechanism of filtration
2) Filter materials

3) Mechanism of regeneration

Design of filter & mechanism of filtration

The most popular way to trap diesel particles is "surface filtration", also known as "wall-flow" filtration. Most commonly, these filters consist of ceramic monoliths with alternating flow channels, that are closed at the ends in order to force the exhaust flow to pass through the porous wall of the honeycomb filtration medium (27,28). In wall-flow filters the solid particles are deposited in the pores and form a layer called "filter cake", which is itself an efficient filter medium. Morphology of soot aggregates that form the filter cake depend strongly from the presence of FBCs and can differ between heavy duty (HD) vehicles and passenger cars (85).

When the exhaust gas is introduced into the DPF wall, the dark gray color around the wall surface becomes darker with time and finally becomes black; the pores open to the channel are defined as 'surface pores'. That is, diesel particulates are at first trapped only inside the surface pores. After the pores are filled up, a transition to surface filtration occurs after and a soot cake if formed. During the process, the thickness of the cake increases continuously, which results in higher filtration efficiency (90% - 99%), in flow restrictions and increasing pressure drop (29,86). Filtration mechanism and a typical wall-flow filter design are illustrated in Figure 2-5 and Figure 2-6 respectively. Pore structure (size, volume, and shape) and filter thickness are responsible for filtration efficiency and also for pressure drop and the materials strength.

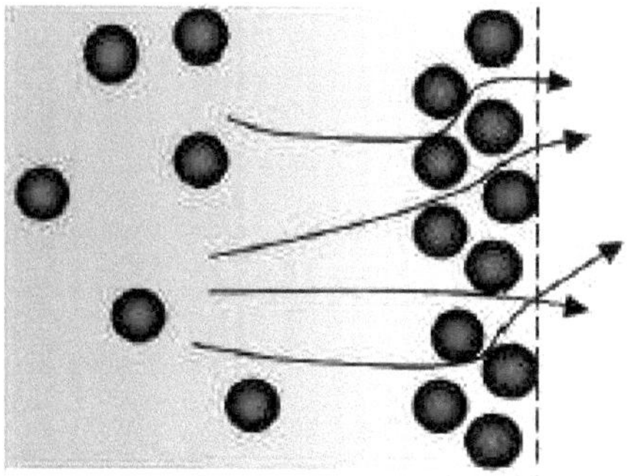

Figure 2-5. Illustration of the flow pattern in a surface filter. The arrows indicate the stream direction of the exhaust gas. The circles represent soot particles which form a cake on the filter surface (29).

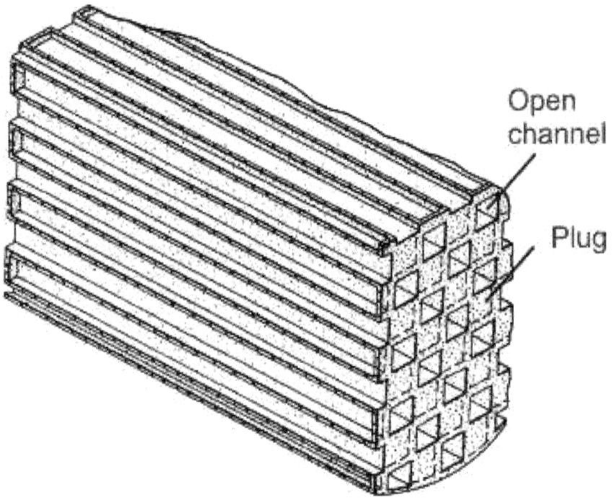

Figure 2-6. Wall-flow (honeycomb) monolith structure. Alternate channels in a ceramic monolith are blocked, forcing exhaust to flow through the porous walls and trapping the soot partiles (87).

The pore size of monoliths made of sintered powder is mainly controlled by powder particle size. A narrow distributed pore size with a mean size in the range of 10 µm - 20 µm at a filter thickness of about 400 µm and a porosity of 50% - 60% are normally the standard specifications for wall-flow filters (83,88).

Other possible DPF designs have a relatively open structure. In such filters particulate matter is found throughout the filter structure and not just on one side of the filter surface, as with wall-flow filter. This mechanism is called deep-bed filtration and it is based on three different ways described below and illustrated in Figure 2-7 (29,44).

Inertial interception: A particle carried along a gas stream, where approaching a collecting body, tends to follow the streamline, but it may well strike an obstruction because of its inertia, becoming removed from the gas-stream. This mechanism is also known as impaction.

Brownian diffusion: Smaller particles, particularly those below 30 nm in diameter, exhibit considerable Brownian movement and do not move uniformly along the gas streamline. These particles diffuse from the gas to the surface of the collecting body and may be collected.

Flow-line interception: If a fluid streamLine passes within one particle radius of the collecting body, a particle traveling along a gas streamLine will touch the body and may be collected without being influenced by inertia or Brownian diffusion (29).

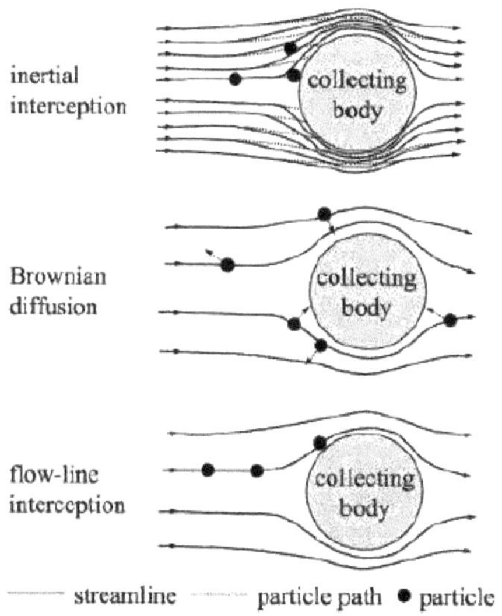

Figure 2-7. Mechanism of filtration in open filter structures (deep-bed filtration) (29).

Deep-bed filtration presents some advantages in comparison to wall-flow filtration as the treatment of relatively high gas flow rates with lower filter pressure and the retention of large soot quantities without dramatic increase of the filter pressure drop or trap thermal failure during regeneration steps. However, in front of these advantages, a generally lower filtration efficiency (typically 30% - 60%) must be accounted (89) and blow-off of the accumulated soot, can also be a problem (87). Despite of that, deep-bed filtration still provides high reduction of particle numbers, in some cases higher than those of wall-flow filters (90). There are many systems that use the deep-bed filtration mechanism, with different design. Among others, PM-Kat® is a system developed by MAN AG. Particulate separation is effected by specially triggered turbulence during the deflection of exhaust gases in the separator and their passage through the sintered metal manifold like the one illustrated in Figure 2-8 (91). It consists of a channelled metal support (coated with a catalyst or with metallic micro-spheres) with metal fiber mats between the individual layers. Deflections inside the channels force parts of the exhaust on top of the fiber mat, which traps the particles (83). PM mass reduction efficiency is 30% - 40% and regeneration occurs by mean of continuously regenerating trap (CRT) principle (see NO_2-aided soot oxidation section). By using this open filter system coupled to an EGR, MAN trucks could achieve the EURO IV emission limits for diesel PM and NO_x emissions (91).

Woven ceramic fibers or metal fibers are also diesel PM deep-bed filters. One typical design involves a layer of continuous fibers wound around a perforated metal tube in such way that the exhaust gas is forced to flow inside the tubes and through the yarn layer, where diesel particulates are deposited on the fibers (87). Generally deep-bed filters are more resistant that wall-flow monoliths, are less damaged from ashes and salts coming from lubrificants and additives and are maintenance-free (87,89,91).

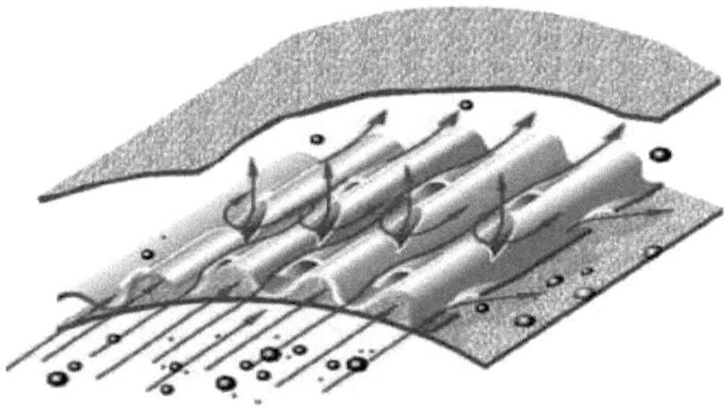

Figure 2-8. Filtration mechanism of an open-filter structure like PM-Kat® (91).

Filter materials

Many different filter materials are available nowadays for construction of DPFs. The material should possess high-temperature stability, resistance against thermal strain, corrosion stability, and sufficient mechanical strength to fulfill all the requirements on filtration, regeneration, and application to ensure an economically and technically sustainable lifetime (83,89). Considering all common material substances, there is no material that covers all these demands perfectly.

Porous cordierite ($Mg_2Al_4Si_5O_{18}$), obtained from clay minerals as caolin and talc, due to its high thermal shock resistance, high filtering efficiency and low costs is the leading candidate material for DPF applications. Pores act as physical traps for PM, and porosity values as high as 50% - 70% by volume are typical for cordierite DPFs. However, the pores, which are essential for the functional performance of the filter, have a deleterious effect on the mechanical properties of the DPF (92). The main material problems for cordierite filters are the low thermal conductivity, which can result in thermal hot spots which exceed the melting point of the material and the predisposal of cordierite to corrosion (66,83).

Silicon carbide (SiC) is a tetrahedral structure of carbon and silicon atoms with strong bonds in the crystal lattice. This produces a very hard and strong material selected for its high filtration efficiency and superior physical properties such as low temperature expansion, high thermal-shock resistance and good thermal conductivity (88). A SiC wall-flow filter was chosen in the first DPF commercially available on a passenger car from Peugeot because of its high resistance to corrosion, originating from the cerium-based FBC (81). The main drawback of SiC filters is its high price (83).

Metals can be either used to construct structures like the PM-Kat® for deep-bed filters or to construct manifolds with surface filtration properties. Porous materials made of high-temperature resistant metals will show similar filtration efficiency when having the same porous structure as their ceramic counterparts. Thus, filters made of sintered metal particles or fibers present a real competition to ceramics (83). The common advantages of metal filters are high strength and high thermal conductivity, resulting in a good thermal shock resistance and a competitive price.

Mechanism of regeneration

A systematic classification of DPF regeneration methods is difficult because of the complexity of the systems and the high number of features. Van Setten *et al.* proposed a classification based on "catalytic" and "non-catalytic" regeneration techniques (29), an approach used also by Neeft and co-workers few years before (43). In this work it has been preferred to divide mechanisms of regeneration into two categories: "active" and "passive" regeneration. These two terms refer to the fact whether or not additional energy has to be provided to oxidize the trapped soot. During active regeneration, somehow additional energy has to be furnished; In passive regeneration, the oxidation reaction is supported by itself. A catalyst can be used in both active and passive regeneration.

In active regeneration the additional energy, used to raise the DPF temperature to accomplish the burning, may come from additional fuel combustion, electric heating elements, enhanced engine-out temperature or microwave irradiation (82).

Additional fuel combustion mainly consist of injecting diesel fuel and pressurized air from a pressure reservoir of the vehicle, introducing them into a combustion chamber (burner), and ignited by means of ignition electrodes. A temperature rise up to 600°C can be achieved. However, this method essentially results in an added fuel consume and increases the complexity of the system in terms of the engine control unit and sensor management (82). Furthermore, excessive heat released during regeneration may lead to a loss of mechanical integrity and catastrophic failure of the filter material (66).

The electric regeneration method involves either the heating of the exhaust gases or the regeneration air by a resistive heater. The heater is placed upstream the DPF where the energy from the heating elements (usually plugs or wires) are deposited to the exhaust stream. Thus, elevating its temperature to the soot ignition temperatures. A combustion front moves through the particulate trap in a downstream direction. The plugs or wires can often be switched off before regeneration has ended, as it propagates spontaneously (43). The temperature within the filter can also be raised internally by incorporating electrical heating elements into the filter structure. In this way, the soot could be ignited directly, without the need to heat all exhaust gas (29). For automotive applications, applying this strategy on-board means additional power consumption for heating the electric resistive element, which in turn has to be drawn from the battery of the car. The energy required to burn the soot is in the range of kilowatts. This puts extra load on the engine and hence results in a relative higher fuel penalty which has been estimated to 5% (43,82).

The engine itself can be also used to produce exhaust gases with elevated temperatures, which are high enough for particulate trap regeneration. For instance, as soon as sensors reveal the necessity to regenerate, the combustion air mass flow rate can be decreased, thus reducing the air-to-fuel ratio. Furthermore, throttling of the intake air or the exhaust gases has been studied and applied. By both methods exhaust gas temperatures can be raised by typically 200°C at medium and high loads. Thus, the trap regenerates in few minutes (43). In-cylinder fuel injection can be also used. Basically, it means that when regeneration is initiated, the engine is operated in the usual way with additional fuel injection at the top end of the cylinder's expansion stroke. There, no additional work can be done and the energy is released as heat (29). Main drawbacks of these techniques are higher fuel consumption, a sensible increase of the NO_x emissions due to the enhanced combustion temperature and the difficulty to regenerate the trap completely at low load and low vehicle speed.

Microwave heating is also an active regeneration method which coupled to filter materials with suitable dielectric properties can relatively easily provide the temperature requested for soot oxidation (80,82). Silicon carbide filters are the most indicated to be regenerated in this way. Despite of that, these systems are not much diffused because of their high battery consumption and complexity.

Almost all these active regeneration techniques can be assisted by the presence of a catalyst which makes a faster soot oxidation possible and/or allows to regenerate a filter at lower temperature. Two general ways have been proposed for the catalytic action on an atomic scale. The so called "electron-transfer mechanism", is thought to act by altering the distribution of π electrons in the graphite sheets, thereby making the carbon substrate more susceptible to oxidation. On the other hand, the proposed "oxygen-transfer mechanism", is considered the general way for the catalytic oxidation reaction. Core of this mechanism are metals which can "oscillate" between two oxidation states thus catalyzing the oxidation of graphite sheets (29). Carbon oxidation is caused by lattice oxygen from the catalyst (reduction step) and re-oxidation of the catalyst by oxygen from the gas phase (oxidation step) (80). Typical catalysts are vanadium, cerium, copper, iron, zirconium, manganese, platinum and palladium (29,43,78).

One problem to solve regarding catalytic regeneration is to create a contact between soot and catalyst. A good way to do that is the use of a fuel-born catalyst. The catalyst is incorporated during the in-cylinder soot-formation process by blending a stable organometallic additive into the fuel (typically 10 – 100 ppm) (81). The catalyst is contained in a specific tank that has to be periodically re-filled. In subsequent evolutions, other automakers chose to directly coat the filter with a catalyst instead of using FBC (78).

During passive regeneration, a trap regenerates by itself without the intervention of on-board diagnostic or control systems and without the input of additional energy. Main advantages of passive regeneration are an increased simplicity of the whole system and a minor fuel consumption, making passive regeneration highly attractive. Actually the increased fuel consumption that the active regeneration methods requires, offsets a significant part of the efficiency gap that separates otto-cycle engines from diesel engines (88), making the latter less attractive.

All passive regeneration techniques involve a catalyst that can be either fuel-born, or coated upstream the filtration media. Passive regeneration is a continuous process in which the catalyst brings the rate of soot oxidation in equilibrium with the rate of soot deposition, which causes a constant pressure drop over the filter (29). "Balance temperature" is defined as the temperature at which equilibrium occurs and for FBC it is usually between 350°C and 450°C (29). These temperatures are rather high, making this kind of regeneration suitable expecially for heavy-duty vehicles.

Some catalysts can oxidize soot without having physical contact. They catalyze the formation of a mobile compound that is more reactive than O_2. It is the case of the so called "spill-over" mechanism, where the catalyst can dissociate oxygen and transfer it to the soot particle where it reacts as it was in a non-catalytic reaction (29,80).

However, the most important passive regeneration mechanism, and possibly the most diffused regeneration method at all, is the NO_2-aided soot oxidation. Because of its central importance in this work, it is discussed more in detail in the next dedicated section.

2.2.5 NO_2-aided soot oxidation

As already stated, soot oxidation reaction in the diesel exhaust atmosphere occurs with noticeable rates at temperatures above 500°C – 600°C, which are rarely met under typical operating conditions. Two main oxidant agents are responsible of the oxidation of soot: O_2 and NO_2 with $[O_2]$ = 5% - 10% vol. >> $[NO_2] \approx$ 4 ppm - 30 ppm (43). On the other hand, NO_2 is much more reactive to soot than O_2. At a concentration higher than 100 ppm, NO_2 is able to oxidize soot already at 200°C; such a temperature can be encountered in diesel exhaust during normal driving cycles. However, NO_2 is present in diesel raw exhaust at very low concentrations (5% - 15% of total NO*x*, or less than 30 ppm), which are not sufficient to assure a fast soot oxidation (93,94). The concentration of NO_2 in the exhaust gas entering the filter can be increased by placing upstream of the filter a diesel oxidation catalyst like Pt, which oxidizes NO to NO_2 thus increasing NO_2 mixing ratio to 50% of the total NO_x, in the temperature range of 300°C -350°C (93,94). This is the principle of the CRT, which simultaneously convert CO and HC to CO_2 with efficiency of 90% (95) and couples a diesel oxidation catalyst to a wall-flow DPF.

Also the PM-Kat® system is exclusively regenerated by mean of NO_2. The impact on NO_x emission should be relevant neither for the legislation nor for the environment. Actually emission limits are valid for NO_x and do not specify between NO and NO_2. Furthermore, even if increasing NO_2/NO_x emission ratio have been recently observed (96), NO in the atmosphere will react to NO_2 in short time scales anyway (95). The main technical limitation of CRT is the requirement of sulfur-free diesel fuel, because SO_2 can poison the oxidation catalyst and when oxidized, it adds up significantly to the total particulate mass (97).

A mechanistic study of the oxidation of carbon by NO_2 and O_2 has revealed two distinct oxidation reactions. A direct reaction occurs between NO_2 and the carbon surface along with a cooperative one involving simultaneously O_2 and NO_2 like shown below (98):

$$C + 2\,NO_2 \rightarrow CO_2 + 2\,NO$$

$$C + NO_2 \rightarrow CO + NO$$

$$C + NO_2 + 1/2\,O_2 \rightarrow CO_2 + NO$$

$$C + 1/2\,O_2\,(+NO_2) \rightarrow CO\,(+NO_2)$$

That study also reported that the reaction of carbon with NO_2 and O_2 is enhanced by the presence of water vapour, which is a component of the exhaust gas stream. However, the activation energy of soot oxidation with NO_2 is considerably lower as compared to soot oxidation with O_2 (94).

The consumption of NO_2 in the reaction is compensated with the production of NO, thus leaving the total NO_x emissions unaltered. Concerning the mechanism of soot oxidation, the reaction is first order with respect to NO_2 and an apparent Arrhenius-type reaction rate has been found (93,97,98). The most favorable temperature range is between 330°C and 400°C (93).

Soot oxidation can be enhanced using a catalyst like Pt, however Pt seems to catalyze only the cooperative soot–NO_2–O_2 oxidation reaction, while the direct reaction between NO_2 and soot is not significantly affected (98). It is found that besides the well established catalytic re-oxidation of NO into NO_2, Pt also exerts a catalytic effect on the cooperative reaction by the formation of atomic oxygen over Pt sites, followed by its transfer to the carbon surface. Thus, the presence of a Pt catalyst increases the surface concentration of –C(O) complexes which then react with NO_2 leading to an enhanced carbon consumption (99). This is the case of the so called catalyzed CRD (CCRD), which mounts a diesel oxidation catalyst (for the NO to NO_2 conversion) in front of the filter which is again coated with Pt. This kind of solution offers the best regeneration kinetics with the lowest balance temperature (100) Other types of catalytic DPFs rely only on the Pt-filter coating both for the NO_2 production and for the soot oxidation enhancement, but their efforts are not encouraging because of a difficult ash management (100).

Ash management is an issue for all kind of DPF, but for the catalytic filters in particular. The catalytic layer is blocked and no contact between soot and catalyst is possible. Ash spans a wide range of compositions, basically depending on the fuel type, lubrification oil and additives. Ashes are normally sulfates, phosphates and oxides of calcium, zinc, magnesium, and other metals. They are formed in the engine's combustion chamber from burning of additives in the engine lubricating oil. Ashes may also come from the wear of the engine (29) or simply enter the combustion chamber with the environment air used for the in-cylinder combustion. Inorganic ashes do not burn, thus often remain trapped in the filter structure and can corrode it or reduce the regeneration efficiency. Solution for this problem are again the use of sulfur-free fuel and low ash-lubrification oil (101).

Regarding the oxidation rates of different types of soot by NO_2 and O_2 in a simulated diesel exhaust atmosphere at 375°C, Messerer *et al.* compared four types of model and real diesel soot: light and heavy duty vehicle engine soot, graphite spark discharge (GfG) soot and hexabenzocoronene (HBC). Experiments showed that up to 75% of carbon mass conversion for HDV exhibited an oxidation rate 3 times higher than for LDV. Moreover, GfG presented in the same experimental conditions, a higher reactivity than diesel soot up to 50% of carbon mass conversion. A correlation between soot reactivity and soot composition led them to suggest that oxidation rate is positively correlated with the oxygen mass fraction and negatively correlated with carbon mass fraction (97,102). Reactivity of soot is a central issue for filter regeneration and it is intensively investigated e.g. by means of innovative methods like Raman-microscopy to determine reactivity structure correlations (103-105). In fact, the tendency of modern diesel engines is to create combustion conditions that produce more reactive soot which would be easily oxidized. That would make the filter regeneration easier.

2.2.6 Secondary effects of DPF regeneration

Besides the desirable result of burning trapped soot, filter regeneration could produce other effects than that. DPFs are reaction chambers where, at high temperatures and in an oxidative atmosphere, soot becomes oxidized. Diesel exhaust represents a complex mixture composed of carbonaceous material, hundreds of combustion products and a great variety of chemical species (43,72,106). Oxidative processes occurring in such an environment could produce compounds other than CO_2 and CO, resulting in secondary emissions of chemical species deriving from soot combustion within the filter. In the last years some attention has been directed to this topic.

From the physical point of view, the size-distribution of the secondary aerosol generated from soot oxidation has been studied. The generation of a significant amount of ultrafine particles (<120 nm) was observed (107). In that study, the filter was regenerated by fuel injection upstream the filter. The formation of these secondary particles was addressed to a series of mechanisms like oxidative fragmentation of soot aggregates, formation of sulfuric acid droplets, condensation of hydrocarbons released or newly generated during the oxidation process. Other studies focusing on unregulated emission from regeneration processes, found similar results (108) or even smaller particles (5 nm - 10 nm)(109). In the first case, the active regeneration was enhanced by in-cylinder retarded fuel injection, thus producing higher amount of HC and CO. The increase in particle number during the regeneration mode, in comparison to filtration mode, was principally addressed to the increased HC emissions caused by delayed fuel injection. Nevertheless both studies concluded that the smaller particle size indicated that the emitted particles were mostly nucleation particles created from the condensation of HC or other volatiles rather than composed from solid carbonaceous matter. To summarize, wall-flow DPFs assure extremely low PM emission for diesel cars during filtration mode, even lower than gasoline cars (110). However higher emissions were observed during short periods of DPF regeneration (109) and immediately afterward, when a soot cake was not yet formed on the filter surface. Beside that, the influence of DPFs on the emitted particles size distribution remains an issue.

Regarding the exhaust chemical composition with focus on the DPFs effect, few literature is available. Diesel particulate filters, like any other catalytic exhaust gas treatment system, are chemical reactors which in principle may also induce the formation of new pollutants (111). The effect of copper- and iron- catalyzed DPFs on the emission pattern of polychlorinated dibenzodioxins/furans (PCDD/Fs) was investigated. While iron-catalyzed DPF did not show any effect, the copper-catalyzed trap clearly changed pattern and increased emission levels inducing an intense PCDD/Fs formation. The formation of lower chlorinated PCDD/Fs was more prominent, resulting in an over-proportional increase of the most toxic tetra- and penta-chlorinated DD/Fs congeners (111).

An issue of major concern and of central importance in this work, is the emission profile of PAH and NPAH. DPFs have a positive effect on PAH emission (35,38), considerably reducing the total PAH emission. On the other hand the emission of NPAH is, at least for certain compounds, increased. First results in this direction were observed with a PM-Kat® used as aftertreatment system for a MAN heavy duty engine (112). Both increase of NPAH concentration and formation of new compounds were observed. To notice is the increase of highly mutagenic 6-NBaP by a factor of 40. Despite of that, the author addressed 20% of this increase to artifact NPAH formation on the sampling filter.

Other studies revealed that DPFs not only supported the formation of certain NPAHs, they also enhanced the degradation of others stressing that the overall contributions of these competing reactions must depend on the operating conditions (38). Two years later the same authors compared NPAH emission of 11 DPF systems (35). They concluded that "most DPFs supported the formation of certain nitro-PAHs but converted others. Therefore, one cannot exclude that some additional toxic compounds not studied herein were formed". Since NPAH are by far more mutagenic and toxic than their parent PAH, the conversion of PAH into NPAH within a DPF would result in a possible increase in the diesel exhaust toxicity thus de facto offsetting the benefices of the diesel trap. The result and discussion section of this dissertation, will give a substantial contribution to the topic, while the next section will give an overview on NPAH formation, occurrence and toxicity.

2.3 Nitrated polycyclic aromatic hydrocarbons (NPAH)

2.3.1 Occurrence and toxicity

NPAH are an ubiquitous class of pollutants known for their high mutagenicity and carcinogenicity. They can be formed via electrophilic nitration during combustion processes, radical reactions or other atmospheric reactions. NPAH are a significant transformation pathway of PAH, but they have a more pronounced mutagenicity than their parent molecules (113).

NPAH, which are formed during combustion via electrophilic substitution in the presence of NO_2 (e.g. 1-nitropyrene and 3-nitrofluoranthene) (38,114,115), are different from those formed by the reaction of PAH with atmospheric OH or NO_3 reactions, which yield 2-nitropyrene (2-NPYR) and 2-nitrofluoranthene (2-NFLT) (116-118). This is because the OH radical or NO_3 radicals attack electron-rich positions first (e.g. 1- position for pyrene and 3- position for fluoranthene), and the subsequent NO_2 addition occurs on the neighboring carbon (119). Quantification of the relative contribution of the specific NPAH isomers may indicate the importance of atmospheric formation processes versus direct emission sources such as diesel engines. That is the case of 1-NPYR and 1,3-, 1,6- and 1,8-dinitropyrenes, whose concentrations are generally much higher in non-aged automobile exhaust particulates than in airborne particulates and 2-NPYR and 2-NFLT which were observed in airborne particulates, but not in non-aged diesel soot (120). An other study concerning NPAH analysis in traffic and background sites showed that the 2-NPYR and 2-NFLT concentrations did not change substantially with sampling location, but the 1-NPYR concentrations were much higher in samples collected at the road-side as compared with sampling locations that were more removed from traffic (121).

NPAHs are thought to account for over 50% of the total vapor- and particle-phase direct mutagenicity of ambient air and particulate matter measured with Ames tests (112). NPAH concentration is normally higher in ultrafine aerosol particle in comparison to fine and coarse particles and they contribute substantially to the mutagenicity of the ultrafine particles (122). Other mutagenicity tests of airborne PM were carried out on *Salmonella* and showed that the

NPAH fraction has the highest mutagenic activity both in comparison to the PAH and the oxygenated-PAH fraction in roadway and park samples. (123).

Among NPAHs produced in diesel engines, 1-NPYR can form DNA adducts, while 1,8- and 1,6-dinitropyrene are the most mutagenic nitroarenes known (124,125) and often called "super-mutagens". They also have carcinogenic potentials (126) higher than those of benzo[a]pyrene or dibenzo[a,l]pyrene, which have the highest toxic equivalent factors among all the native PAHs (127,128). Almost all NPAH can directly be mutagenic to DNA while others need first to be metabolized. The orientation of the nitro-group is a critical structural feature that can affect mutagenicity and carcinogenicity of NPAHs (129). For instance 1- and 3-nitrobenzo(a)pyrene (1- and 3- NBaP) are strong direct mutagens while 6-nitrobenzo(a)pyrene (6-NBaP) needs first a metabolic activation (130). Also the DNA adducts formed are substantially different and often are NPAH-specific (131).

For these reasons NPAH are a class of pollutants of great concern for public health which, even in absence of maximum tolerated concentrations or limits of emissions defined by the legislator, becomes constantly monitored from scientists all over the world. The next section briefly review the most common analytical procedures for the identification and quantification of NPAHs.

2.3.2 Analysis of PAH and NPAH

The great majority of the analytical techniques employed for PAH and NPAH analysis requires the extraction of the substance from its matrix (e.g. water, soot, aerosol, soil, food etc.) and the clean-up of the extracts prior to analysis (119,132). Soxhlet extraction, sonication and pressurized fluid extraction are the most conventional extraction methods (119,133). Internal standards (I.S.) can be added to correct final data for incomplete extraction or losses during analysis. In case of mass spectrometric quantification, deuterated PAH and NPAH are used as I .S. (119).

Because of the molecular similarity of PAH and NPAH, most of the analytical methods nowadays available for PAH determination are also suitable for NPAH. Nevertheless, in environmental samples or in complex matrices like diesel soot, NPAH concentrations are usually 10 – 100 times lower than those of PAH, often reaching the detection limit of many analytical instruments (128). Gas chromatography (GC) is the most spread tool for the of analysis of volatile PAH and NPAH because of its high separation efficiency of complex non-polar or light-polar analytes (134). As detection method, mass spectrometry (MS) is often utilized because of its ability of recording "fingerprint" mass spectra of individual components (119). The most common preferred MS analysis methods consist of electron impact, negative ion chemical ionization or atmospheric pressure chemical ionization (135-138) coupled to selective ion monitoring, time of flight or quadrople mass spectrometers (112,136,139).

High-performance liquid chromatography (HPLC) is a valuable alternative to GC due to its high resolution and simplicity (119). HPLC also offers the possibility to collect separated eluate fractions, which can be used for a second separation (thus obtaining a better resolution) (132,140) or for other purposes (e.g toxicological tests). HPLC with chemiluminescence or fluorescence detection have been optimized and represent routine analysis methods. Their limit of detection are in the low pg range both for PAH and NPAH. Nevertheless, to analyze NAPH, both applications need to reduce NPAH to the corresponding amino-PAH to enhance their chemiluminescence or fluorescence properties (112,141). The on-line reduction is the preferred method and is currently achieved with Pt-coated alumina columns, Pt/Rh-coated alumina and zinc/glass bead packed columns (112,142). HPLC can be also used in combination with MS and, even if the sensitivity of this technique can be lower than that of fluorescence or chemiluminescence detection-, unknown species can be identified offering a significant advantage (119,143,144).

Immuno-assays are class of analytical methods that should be mentioned in regard to analysis of PAH and NPAH. For instance, the occupational exposure to diesel exhaust was evaluated by analyzing PAH and NPAH urinary metabolites by means of a competitive enzyme-linked immunosorbent assay (ELISA) (145). Other studies showed that ELISA could be used to detect 1-NPYR in particulate matter. However, in that case the results obtained from the immuno assay where about six times higher than those delivered from the GC-MS analysis used for validation (146). That was mainly due to the cross-reactivity with 2-nitropyrene and

2-nitrofluoranthene. Cross-reactivity is actually one of the main limitations of the immuno assays for PAH and NPAH (146,147). Hence, they are suitable mainly for semi-quantitative analysis of PAH and NPAH(148). On the other hand, if specific antibodies are available, immuno assays offer advantages in terms of costs and time-consumption, making them attractive for continuative monitoring and screening tests.

2.4 Biofuels

2.4.1 Biofuels: occurrence and production

During the last years the interest in alternative fuels for the automotive sector has raised significantly. At present biofuels are one of the most established and widespread technology as an alternative to common fossil fuels. The European Commission adopted in 2003 a directive in which was determined that member states must ensure that the minimum share of biofuels sold on their markets had to be 2% by December 31st 2005, and will have to reach 5.75% by December 2010 (149) .

The term biofuel generically defines a fuel derived from biomass material, but the variety of raw materials, production processes and final products is considerable (150-152).

Vegetable oil is the oldest diesel fuel known. Rudolf Diesel (*1858 – † 1913), the inventor of the engine that bears his name, used peanut oil as fuel in his invention (149,150). Nevertheless, the use of vegetable oil in the direct-injection type diesel engine is limited by their high viscosity, which is nearly 10 times higher than that of fossil diesel fuel (153). To overcome this problem, engine modification like dual fueling or heated fuel lines can be adopted. Also fuel modifications like pyrolysis or micro-emulsion reduce viscosity. However, transesterification appears to be the most promising solution (150). Chemically, transesterification (also called alcoholysis) means taking a triglyceride molecule (*i.e.* vegetable oil) and removing the glycerine by substituting with an alcohol (*i.e.* methanol) (154). Fatty acid methyl ester (FAME), also known as biodiesel, is the product of this process.

The usual raw materials, being exploited commercially to produce biodiesel, consist of edible fatty oils derived from rapeseed, soybean, palm, sunflower, and other plants (152). Attempts to produce biodiesel from non-edible products as algae (155) or waste frying oil (54,156) have also been done with encouraging results. Biodiesel can be blended in any proportion with mineral diesel to create a stable biodiesel blend. The level of blending with fossil diesel is referred as Bxx, where xx indicates the amount of biodiesel in the blend (*i.e.* B10 blend is 10% biodiesel and

90% diesel). It can be used in diesel engines with no major modification of the engine hardware (157).

Concerning fuels for otto-cycle engines, ethanol has been known as a fuel for many decades since it has an higher octane number than gasoline (157). It is mainly produced by fermentation of sugar or starch which again is obtained from cultivation of sugarcane, sugar beet or maize. The so called "second generation" ethanol can be obtained from lignocellulosic biomass which represent abundant feedstock for ethanol production and does not compete for use as food. For this reason plants like *giant miscanthus* are cultivated (151). Blends of ethanol in gasoline are commonly used in otto-cycle engines; however, modifications at the compression rate, carburetion, fuel lines and valve housings are required for alcohol fueling (157).

2.4.2 Biofuels: emissions and health

Since many years the composition of vehicles emissions is controlled. For some pollutants limits of emission (LOE) are specified by legislation. Regulated pollutants are CO, HC, nitrogen oxides NO_x and PM. In Europe, for each pollutant three different LOE are set for passenger cars, heavy duty vehicles and off-road vehicles respectively. Other pollutants PAH, their nitrated derivatives NPAH or aldehydes, are called unregulated pollutants, as no LOE has been defined

The high oxygen content in biodiesel and ethanol results in the improvement of its burning efficiency, thus a positive effect on vehicle emissions could be expected. It has already been found that blending fossil diesel fuels with biodiesel can reduce the emission of all regulated pollutants with exception of NO_x, which shows an increase (158-162). In otto-cycle engines, the effect of ethanol addition to gasoline on pollutant formation is mainly that PM and CO emissions get reduced, while NO_x emissions are not significantly influenced (157,163,164).

Regarding unregulated emissions, it is not simple to determine the influence of biofuels. The reason is the great number and diversity of compounds, which can be emitted and their emission can individually be affected by the use of biofuels. Many authors report a decrease in the diesel engines PAH emission by using biodiesel (104,161,165-167), but others found no statistically significant differences (168,169) or even an increase (54). Differences can be addressed to the high variety of fuels, test cycles and vehicles used in these studies. On the other hand, there is a lack of data on NPAH emission. Regarding diesel engines, in two cases it is reported that concentration of NPAH were decreased with the use of biodiesel (161,170) but other authors found no significant differences between conventional fuel and biodiesel blends (169). To our knowledge no literature data are available for PAH and NPAH emission in otto-cycle engines using ethanol or fossil gasoline/ethanol blends.

Aldehyde compounds are also unregulated pollutants emitted from engines. The most emitted compounds are formaldehyde and acetaldehyde. Some authors report an increase of their emission in diesel engines operated with blends with a low percentage of Biodiesel (B2, B5 and B10), but a decrease for B20 and B50 (171). Adding biodiesel can lead to lower emissions of carbonyls (167) nevertheless, also negative effects were observed (161). Blending fossil diesel fuel with 15% ethanol resulted in higher emission of many carbonyl compounds (4). In otto-cycle engines, the main effects of ethanol blending were increased acetaldehyde emissions, and reduced emissions of all other carbonyls except formaldehyde and acrolein (95).

Due to the fact that many unregulated pollutant are classified as mutagenic and/or carcinogenic, differences in their emission due to use of biofuel can influence the toxicity and/or the mutagenicity and/or the carcinogenic potential of vehicular emissions. The debate on the topic is quite intense and results are not univocal. Biodiesel PM was found to be more toxic than diesel because it promoted cardiovascular alterations as well as pulmonary and systemic inflammation (172). It has been reported that rapeseed oil emission caused stronger mutagenicity than the reference diesel fuel. In the same studies, biodiesel emission had a moderate, but significantly higher mutagenic response in assays of *Salmonella* strain TA98 with metabolic activation and of *Salmonella* strain TA100 without metabolic activation (149,173). On the other hand, other authors report that the addition of 20% biodiesel to fossil fuel did not exert any remarkable effect on total mutagenicity and mutagenic profile of the emission (169).

3 Experimental

3.1 Aerosol and gas mixtures generation

The use of diesel engines and/or real diesel soot for test experiments can be complicated. It requires expensive test benches and proper personal. Moreover the constancy of such systems can be a problem as well as the control and the change of single parameters like temperature, NO_2 concentration, oxygen and water content, etc. Therefore the first approach of this work was to use artificial soot as proxy for real diesel soot under controlled laboratory conditions.

Diesel exhaust aerosol was simulated by coating artificial soot with PYR or BaP and adding O_2, water vapor and NO_2. The aim of this section is to illustrate how the simulated diesel exhaust aerosol was produced.

3.1.1 Spark discharge soot

In many studies, spark discharge soot (GfG) soot has been used as model soot. In this work it has been chosen because it offers several advantages:

- It is easy to produce
- Its production is constant over a long time
- Particle number concentration and size distribution are in the same range of that of diesel exhaust soot

The aerosol generator (GfG 1000, Palas, Karlsruhe, Germany) generates particle agglomerates from pure graphite. Soot is produced by a spark discharge between two graphite electrodes in argon as carrier gas. The electrodes, both 6 mm in diameter, are mounted in brass collets. The flat ends of the electrodes are positioned at a distance of about 2 mm. A high-voltage capacitor with a capacitance of 20 nF, which is connected to one of the electrodes, is charged by a high-voltage supply with adjustable output current.

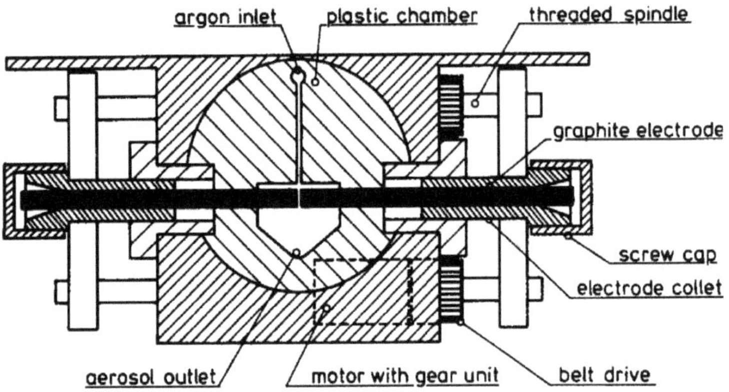

Figure 3-1: Cross-sectional drawing of a Palas GfG 1000 soot generator (174).

When the breakdown voltage (2 kV) is reached, it discharges rapidly in a spark via the electrode gap (174). Aerosol mass produced is proportional to the spark frequency that is adjustable in a range of 2 Hz to 300 Hz.

The resulting vaporised graphite condenses homogeneously to very fine particles of pure carbon which coagulate to agglomerates during further transport or residence time. The primary particles are 5 nm – 10 nm in diameter. The mean size of the chain-like agglomerates is usually between 20 nm and 200 nm depending on the operating conditions.

3.1.2 PAH coating system

The argon carrier gas (7 L min^{-1}, 99.995% Ar, Messer-Griesheim, Germany) containing the soot aerosol produced by the spark discharge soot generator was fed into an agglomeration flask (0,5 L three-neck glass flask) that also served as a buffer reservoir. A separate flow of nitrogen (1 L min^{-1}, 99.999%, Messer-Griesheim, Germany) was allowed to pass over solid PAH (2 g of PYR or 0.5 g of BaP) that was placed in a temperature controlled reservoir. The soot containing argon gas was then mixed with the PYR or BaP containing nitrogen gas in a ring-gap mixing nozzle and subsequently passed into a condenser set at 10°C. According to previous findings

(175) this procedure let PAH to become adsorbed onto the soot particles. This setup is depicted in Figure 3-2 and has already been tested and published (176). To determine the size distribution of PAH-coated particles, 0.3 L min^{-1} of the total 9 L min^{-1} flow were directed to a scanning mobility particle sizer (SMPS) composed of an electrostatic classifier (model 3071, TSI, Aachen, Germany) and a condensation particle counter (CPC) (model 3025 TSI, Aachen, Germany). Assisted by a vacuum pump, the remaining 7.7 L min^{-1} aerosol containing flow was collected on a quartz fiber filter (QR 100, 47 mm, Advantec, Japan). Prior to use, the filters were heated at 500°C for 12 h to remove organic contaminants.

For gravimetric mass determination of the sampled soot, filters were weighed before and after collection by first equilibrating them for temperature and relative humidity in a glass desiccator filled with silica gel (Merck, Darmstadt, Germany). The standard sampling time was 1 h and the average mass deposited on the filter was 0.47 ± 0.10mg (n = 60).

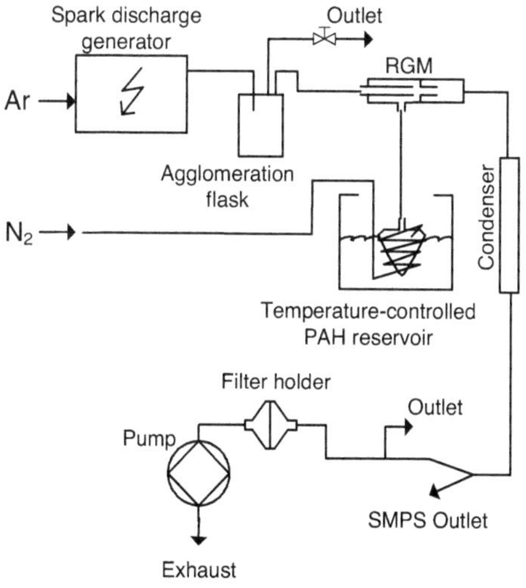

Figure 3-2. Aerosol generation, coating section, sampling system (V: valve; RGM: ring gap mixing nozzle).

3.1.3 PAH combustion experiments

For PAH aerosol combustion experiments it was at first necessary to produce a pure PAH aerosol and to collect it on a quartz fiber filter. Perylene is a five-ring PAH with a melting point of 274°C. It was chosen for its similarity to PYR and BaP and for its low toxicity (177). The PAH aerosol generator system is illustrated in Figure 3-3. It consists of a small ceramic tray containing 8 mg of perylene which was positioned in the middle of a quartz tube (1.5 m long, ϕ = 12 mm). The quartz tube was inserted in a housing of quartz (40 cm; ϕ = 30 mm), which was enveloped by a heating band (Hillesheim HSS/040, Germany).

Temperature was controlled by a temperature sensor (HKMTSS-IM150U-150, Omega, Newport, UK) positioned between the housing and the quartz tube. The end of the tube was connected to a filter holder and the particle concentration was measured on-line by a condensation nuclei counter (CNC 3020; TSI). A flow of N_2 of 1.5 L min^{-1} was applied and the all sampling was assisted by a pump. Perylene aerosol particle production started at 250°C. In about 20 minutes at 300°C all perylene sublimated from the ceramic tray. About 50% of the perylene was collected on the quartz-fiber filter positioned in the filter holder. The rest was condensed at the tubing wall because of the temperature drop outside the quartz housing.

The quartz fiber filter was then transferred to the temperature programmed oxidation (TPO) test bench. This test bench has already been described elsewhere in detail (15,178). Briefly, it offers the possibility to oxidize samples under a controlled atmosphere and to follow the oxidation process by mean of a FTIR spectrometer in real time.

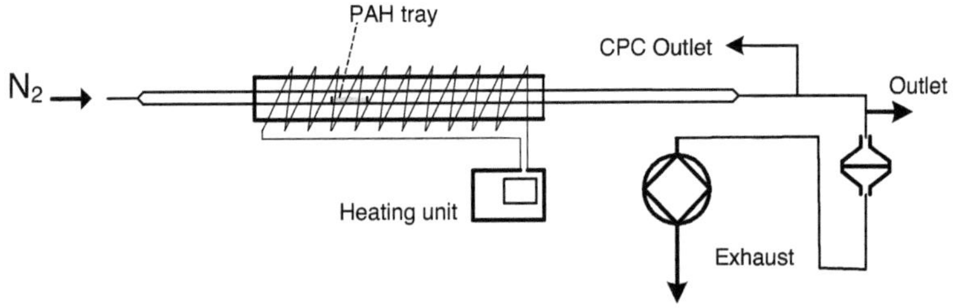

Figure 3-3. PAH aerosol generator system.

3.1.4 Nitrogen dioxide

Known amounts of NO_2 were generated by means of two permeation tubes (179). They consist of two small flasks (permeation tube A: glass made, 8 mL; permeation tube B: steel made, 0.5 mL) tightly closed with PTFE membranes (Teflon®). Both were filled with condensed NO_2 which permeates through the membrane in gas form at a constant rate when the tubes are held at constant temperature. By passing nitrogen over the permeation tubes, a wide range of concentrations can be obtained by varying the flow rate of the carrier gas, the temperature, the surface of the permeation membrane, or its thickness (179). The permeation tubes were thermostated at 37°C, and the PTFE membranes thickness was 1 mm and 4 mm for permeation tube A and B, respectively.

Permeation rates were determined by weighing the tubes at different time intervals and by calculating the amount of NO_2 that has permeated. They were A = 5.37 µg NO_2 min^{-1} and B = 0.19 µg NO_2 min^{-1}. The calculated pressure within the permeation tubes was estimated to be approx. 2 bar. Tube A was used for concentration betweens 6 ppm (m/m) and 1 ppm. Tube B for concentrations below 1 ppm. With such a rate, permeation tube A was filled only once an lasted for two years and half. Tube B has to be refilled approx. every 16 months.

To obtain higher NO_2 concentrations, a bottle with 1,018% (molar ratio) of NO_2 in synthetic air (15 L, Messer-Griesheim, Germany) was used. By means of a pressure regulator (EN 562, Messer, Germany) a volume down to 20 mL min^{-1} could be set, therefore it was possible to achieve NO_2 concentrations in the range of 20 ppm - 200 ppm.

3.2 Aerosol and gas mixture control

The produced aerosol and gas mixtures had to be characterized. Aerosol size distribution, aerosol particle concentration as well as NO_2 concentration could be monitored by on-line or off-line methods. This section describes the systems used to accomplish that and provides a briefly review of the theory supporting the functioning of these systems.

3.2.1 Scanning mobility particle sizer (SMPS)

The aerosol production and coating system was prepared with several outlets which, if opened, directed the aerosol to a SMPS. In this way, the on-line monitoring of the aerosol size distribution and particle concentration at different points (e.g. before PAH coating, after PAH coating, before DPF, after DPF etc.) was possible. A SMPS system is composed of two main instruments: a differential mobility analyzer (DMA) and a CPC.

A DMA classifies aerosol particles according to their electrical mobility *Z* which for spherical particles is given by

$$Z = \frac{neC}{3\pi\mu D_p},$$

where *n* is the number of elementary charges carried by the particle, *e* is the magnitude of the elementary unit of charge, *C* is the slip correction factor (again a function of particle diameter D_p, and pressure P) and μ is the absolute gas viscosity. *Z* depends on gas properties, particle charge, and the geometric particle size but is independent of other particle properties such as density (180). For instance, the electrical mobility of a singly charged particle is a monotonically decreasing function of particle size, but two particles with the same diameter can have different mobilities, if they carry a different charge. Therefore, since a DMA classifies aerosol particles according to *Z*, the result of a measurement is a mobility distribution (i.e. number of particles that have a certain electrical mobility). One can derive the size distribution (i.e. number of particles that have a certain diameter) only if the charge distribution of the aerosol is known.

For this reason, the DMA used was equipped with a radioactive bipolar charger or "neutralizer" in form of a thin wall stainless steel tube containing 2 millicuries of the radioactive gas ^{85}Kr. The β-radiation produced by the ^{85}Kr source can penetrate the thin wall and ionizes the gas molecules outside, producing large quantities of bipolar ions (ions of both polarities). The aerosol particles are then directed to this highly ionized environment and discharge themselves by capturing ions of opposite polarity. Within a time depending on the ion concentration (typically a few milliseconds), an equilibrium state is attained where the particles carry a bipolar charge distribution given by Boltzmann´s law (181). After passing through the neutralizer, the aerosol is charged with known charge distribution and can enter the DMA.

The analyzer consists of two concentric metal cylinders. The inner cylinder is negatively charged (collector rod) and the outer is grounded. Aerosol and clean air are introduced into the apparatus near the top and flow down the annular space between the cylinders. Particles of positive polarity are attracted to the collector rod. Particles with high Z will impact on the upper portion of the rod and those with lower Z will be carried along with the main outlet flow. Only those particles with the correct electrical mobility will be attracted to the sampling slit resulting in an aerosol composed of particles with the same electrical mobility (181). By scanning the voltage applied within a certain range, the Z requested to enter the sampling slit varies with the voltage, and particles with different electrical mobility can be collected. The DMA used in this work is a TSI 3071 model and is illustrated in Figure 3-4.

If a particle counter is positioned at the outlet of the DMA, it is possible to on-line count the number of particles which have a certain Z thus obtaining an electrical mobility distribution. To calculate the electrical mobility diameter, raw mobility data have to be corrected for the charge distribution probability obtained from the Boltzmann´s law. This diameter must not be confused with the aerodynamic diameter which can be calculated from an electrical mobility diameter by mean of some corrections nicely reported by Peters *et al.* (182).

The counter used was a TSI CPC 3025. In this model, aerosol particles pass through n-butanol vapor which condenses on the particle, forming a droplet. Each droplet is large enough to scatter a detectable amount of light when it passes through a light beam while the dry particle would have been much too small to be detected.

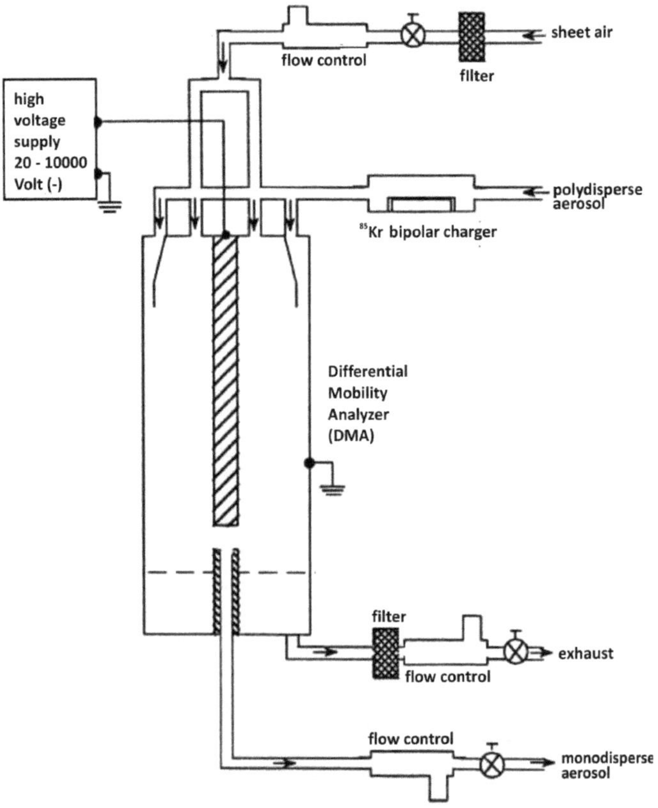

Figure 3-4. Schematic diagram of the TSI 3071 electrostatic classifier (181).

The droplet size is nearly independent on the particle size, therefore the light scattering is only dependent on the droplet concentration and not on the size distribution (180,183).

3.2.2 Photometric mircrodetermination of NO_2

As explained in section 3.1.4 NO_2 concentration was determined by diluting known amounts of NO_2 emitted from permeation tubes or coming out from a test gas bottle. As orthogonal method to confirm the calculated data, the content of nitrogen dioxide was determined also with the analysis described by Saltzman (184).

This method is based on the reaction of NO_2 with an absorbing reagent where sulfanilic acid produces diazosulfanilic acid. The acid reacts with N-(1-Naphthyl)-ethylendiamine to form a violet colorant. Calibration was made with a standard sodium nitrite solution that reacts with sulfanilic acid in the same way.

For NO_2 determination with this method, 30 mL or 50 mL of absorbing reagent (the latter volume for concentration of NO_2 > 10 ppm) were placed in a midget fritted washing flask. Sample was pumped though it with a flow 0.4 L min^{-1} until a sufficient color was developed. For concentration > 20 ppm, often one minute was enough. After the complete color formation (usually 10 min after the sampling) the absorbance was red in an UV-Vis spectrometer at 550 nm.

3.2.3 Chemiluminescence NO/NO_x analyzer

For the on line monitoring of the NO_2 concentration, in the last part of this work also a chemiluminescence NO_x analyzer (CLD 700 Al, Eco Physics, Munich, Germany) was used. It is based on the chemiluminescence reaction of NO with O_3. Ozone is generated internally by passing a silent electrical discharge through the sample. O_3 then reacts with the NO (the actual analyte) to form NO_2 and O_2. About 20% of the total NO_2 formed is in an electronic excited state which returns to ground state by emission of radiation with wavelengths between 600 nm and 3000 nm. This is indeed the chemiluminescence signal which, in the visible range, can be detected by means of a photo-multiplier. In presence of excess of ozone, the signal is proportional only to NO concentration. To detect NO_2, it is necessary to reduce it to NO by passing the sample through a thermal catalytic converter. The reduction agent is molybdenum and at 375°C the following reaction occurs:

$$3NO_2 + Mo \rightarrow 3NO + MoO_3$$

The reaction is not reversible, therefore the max. life-time is approx. 1000 ppmh. Analysis is carried out in two separate chambers where sample is initially equally split. In the first chamber

NO is directly detected, while in the second chamber at first catalytic NO_2 reduction takes place. The sum of the two signals gives the total NO_x concentration and the NO_2 amount is obtained as difference of $NO_x - NO = NO_2$

3.2.4 NO_2 scrubbing by means of annular denuders

Sampling PAH-coated soot from an atmosphere with diesel exhaust NO_2 concentrations, without NPAH artifact formation, is a challenge. Actually the NO_2 could directly nitrate the soot-adsorbed PAH with consequent artifact NPAH formation. To solve this problem it is necessary to dispose of systems that remove the NO_2 from the atmosphere without interfering with the aerosol particles.

Active carbon has a high affinity for NO_x and its uptake potential is already well known (185-187). Actually such material has been already investigated as method for NO_x emission control or as a means to sequester and then concentrate NO_x for increasing the efficiency of catalytic reduction processes, which are not especially suited to diluted gas streams (188). However, since active carbon can adsorb PAH even at low concentrations (189,190), this material was not suitable for the purpose of this work. Therefore, to eliminate the NO_2 from the aerosol flow, annular denuders with a proper coating were used.

In annular denuders, air is forced to pass through the annular spaces between two or more concentric tubes.

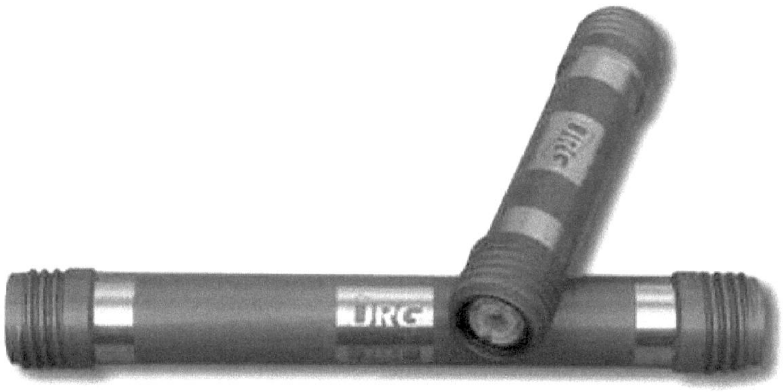

Figure 3-5. Three channels annular denuders used in this study.

By increasing the number of concentric tubes, the length needed for effective scrubbing decreases. The denuder used consisted of four, 20-cm long concentric sand-blast glass cylinders. The cylinders had a diameter of 8 mm, 12 mm, 16 mm and 20 mm, respectively. The tubes had to be coated with a substance that acts as NO_2 sorbent, so that every NO_2 molecule that hit the wall would have been chemisorbed. Two coating solutions were tested: 10% potassium iodide, 5% ethylene glycol and 1% sodium arsenite in methanol (191), and a saturated guiacol/methanol solution.

Wall coating was applied by filling the annular spaces with solution and then emptying them. Following this operation, dry clean nitrogen was passed through the denuder until a smooth wall coating was obtained. The operation could be repeated two or three times to obtain a better coating. Denuders used are illustrated in Figure 3-5.

The measurement of the NO_2 scrubbing efficiency was performed at various flow rates by introducing into the denuder known amounts of NO_2 in the 1 ppm - 120 ppm (m/m) concentration range and measuring the outlet concentration with the chemiluminescence NO_x analyzer or with the Saltzman method explained in detail in precedent sections.

3.3 Nitration experiments

3.3.1 Quartz fiber filter

For the step of exposing the PAH-coated soot to NO_2, the filters were placed into a specially designed stainless steel heatable filter holder through which NO_2 containing gas was pumped as shown in Figure 3-6.

The background gas comprised 5% O_2 and 95% N_2, as analogous conditions were used successfully in studies simulating diesel exhaust conditions (105,192,193). The background gas was saturated at 100% relative humidity prior to mixing with NO_2 by passing it into a chamber of boiling bi-distilled water and cooling it to 22°C with a condenser (162). With a ring-gap mixing nozzle the water saturated nitrogen/oxygen flow was then mixed with the nitrogen flow carrying the NO_2 to obtain the desired NO_2 concentration.

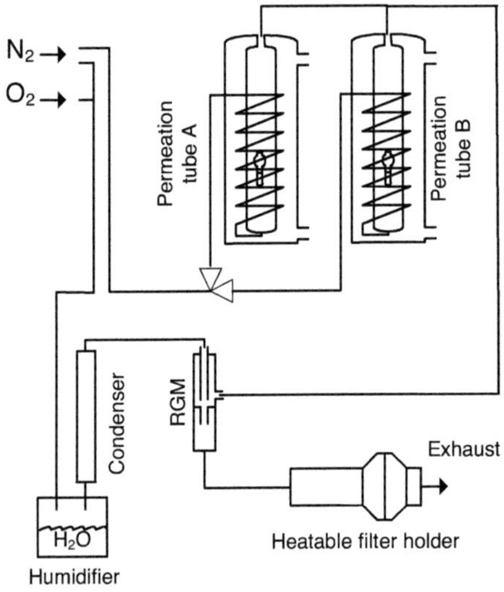

Figure 3-6. Soot nitration system.

Reproducibility tests were carried out by cutting 15 loaded filters into two equal parts (part A and part B) using a ceramic knife. Each part was re-weighed to determine the amount of soot on each. Reproducibility of the analysis was estimated by nitrating both parts of the same filter simultaneously (part A_n and part B_n). The effect of a different filter loading or coating was estimated by using each part of two different filters simultaneously (part A_n and part A_m). The filter holder used for nitration was connected to a heating unit (HT52 Hillesheim, Germany) to perform experiments at different temperatures ranging from 20°C up to 300°C.

For experiments on the gas/solid phase partitioning of evolved NPAHs, quartz-fiber filter were spiked with known amounts of 1-NPYR and 6-NBaP, inserted in the heatable filter holder and left under a flow of N_2 of 1.5 L min^{-1} at different temperatures for 1 h. For collection of evolved gases two polyurethane foam plugs (PUF) (GA3035, Ziemer Chromatographie, Langerwehe, Germany) were inserted in line at the outlet of the filter holder in an airtight compartment.

Gas-phase sampling was possible also during nitration experiments and the PUF collecting system was compared with a midget fritted washing flask containing 100 mL of toluene. Both systems showed comparable collection efficiencies (see section 4.2.3).

3.3.2 PM-Kat® simulation system

Spark discharge soot particles were coated with BaP in the same way as described in section 3.1.2. The aerosol was then passed through a heatable flat bed reactor (FBR) already described in detail elsewhere (194). It could be heated up to 300°C by mean of 5 heating cartridges (firerod 10 mm x 50 mm, 150W, 19.46 W/cm^2, Watlow GmbH, Kronau, Germany). In the 350-mm long flow channel of the FBR, different structures for the deposition of particles can be inserted. To simulate a real DPF with open structure (e.g. PM-Kat®) typical for heavy duty vehicles, 2 corrugated stainless steel structures (Oberland Mangold, Eschenlohe, Germany) of 74 mm length (cell density ≈ 160 cpsi) coated with steel micro spheres (diameters of 50 - 150 µm) were positioned in the middle of the flow channel. The total number concentration particle removal efficiency of the system, determined by SMPS measurements, resulted in approx. 40%. Just before entering the FBR, the aerosol containing gas was mixed with the NO_2 provided from a gas bottle (60 cm^3 min^{-1}, 1.018% NO_2 in N_2, Messer-Griesheim, Germany). The NO_2 concentration in the flow channel was 120 ppm. The experiment duration was 1 hour. At the outlet of the FBR the gas flow was split and 1.5 L min^{-1} were directed to a quartz fiber filter to collect both the aerosol particles that were not intercepted from the open filter structures and those that were eventually re-emitted after the deep bed filtration.

To avoid further nitration of the BaP collected on the quartz fiber filter with the consequent formation of artifacts, it was necessary to drastically reduce the NO_2 concentration in the aerosol prior to filtration. To achieve that, two parallel annular denuders were placed in front of the filter holder. A complete denuder description and characterization is provided in section 3.2.4. The efficiency of this denuder system was tested by measuring at regular time intervals the [NO_2] after the denuder with the Saltzman method (184). After 10 min it was > 98% and after 1 h it was still > 90%. To further reduce the [NO_2], the aerosol was diluted downstream the scrubbing with 6 L min^{-1} of N_2. After 1 hour of sampling the [NO_2] was < 0.2 ppm. The gas phase was collected by a fritted washing flask filled with 150 mL of toluene that was placed directly after the filter holder.

This set-up is described in Figure 3-7. The set-up allows to obtain artifact-free samples at three points: a) DPF structure; b) quartz fiber filter after DPF; c) washing flask for the gas phase collection.

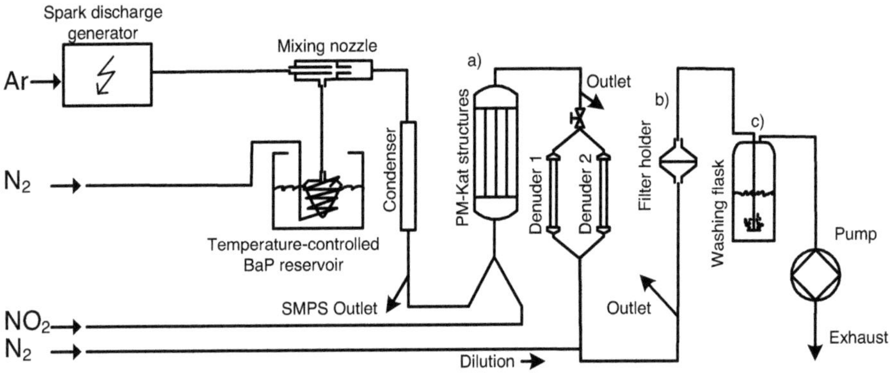

Figure 3-7. DPF-simulation system (PM-Kat®).

3.3.3 DPF simulation system

The aim of the DPF simulation system was to reproduce in the laboratory conditions similar to those encountered in a real wall-flow particulate filter. Therefore a cordierite honeycomb DPF (AC 200 cpsi, Duratrap) was placed in a home-made, air-tight stainless steel housing (42 mm x 2 mm x 145 mm).

The BaP coated aerosol, the NO_2 containing gas and the 5% oxygen (previously saturated with water vapor) were mixed just in front of the FBR (without PM-Kat® structures) which served as pre-heater (200°C or 300°C) for the DPF. The DPF itself was again heated by means of a heating tape (HSS/010 Hillesheim, Germany) and its temperature also depending on the temperature set in the pre-heater could be adjusted between 20°C and 250°C.

The sampling part of this system was the same as in the PM-Kat® simulation system. Briefly, the DPF-outlet aerosol was split and 7 L min^{-1} were lead out of the system and the remaining 1.5 L min^{-1} were first directed to the denuders, then diluted with 6 L min^{-1} of nitrogen and finally collected on a quartz fiber filter (particulate phase) and in a midget fritted washing flask filled with 100 mL of toluene (gas phase). Because it was not possible to remove the DPF from its housing after every run, only the quartz fiber filter and the toluene contained in the washing flask could be analyzed.

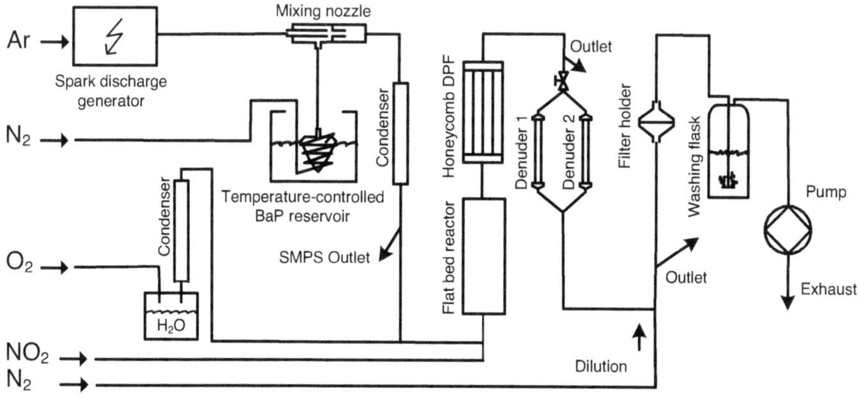

Figure 3-8. DPF simulation system (wall-flow).

3.4 Vehicle exhaust emission sampling

3.4.1 Biofuel project: sampling at TU Wien and TU Graz

Real exhaust soot sampling at the rolling test bench of TU Vienna and TU Graz was requested for the analysis of soot-adsorbed NPAH, within the biofuels project. For this purpose quartz fiber filters (QR-100, ϕ 70 mm, Advantec, Japan) were used. Prior to use, the filters were heated at 500°C for 12 h to remove organic contaminants. Both rolling test benches were equipped with a constant volume sampler (CVS) dilution tunnel. The dilution tunnel temperature was set at 52°C. From the dilution tunnel, by means of a counter-current isoaxial-sampling system, 90 L min^{-1} were directed to 3 filters in parallel (30 L min^{-1} per filter). The sampling volume per filter was regulated by means of critical orifices and the total sampling volume was controlled by a volume-counter (Elster GmbH, Germany) (15). Sampling duration time was dependent on the driving cycle used for the test. After the sampling, the filter were covered with aluminum foil and kept at a temperature of –7°C until laboratory analysis.

3.4.1.1 Vehicles, fuels and test cycles

A heavy duty vehicle (MAN TGA 24.440), two passenger cars (Saab 95 2.3 T Biopower; Audi A4 Avant 2.0 TDI DPF) and an off-road engine (John Deere 6068) were used during the tests. Details on vehicles and engine specifications are illustrated in Table 3-1. Test vehicles and engines were operated with fossil fuels, biofuels as well as mixtures with different amounts of biofuels in fossil fuel. Biofuels were biodiesel (FAME), ethanol and vegetable oil. The level of blending with fossil fuel is referred to as Bxx or Exx, where xx indicates the amount of biodiesel or ethanol in the blend. The test cycles were representative of the actual legislation. Test cycles are different for each class of vehicles and can be transient or stationary. Transient cycles simulate the driving under a real context (urban, extra-urban) with alternate acceleration, deceleration and idle periods. In stationary cycles engines are tested on an engine dynamometer over a sequence of 10 - 13 steady-state modes with different speed and load. Each mode has to be held for a determined period (2 to 4 minutes).

Table 3-1. Engine types, aftertreatment system and emission levels of the vehicles used in the biofuel project.

Test vehicle	Engine	Exhaust aftertreatment	Emission levels
Heavy Duty Vehicle (MAN TGA, diesel)	6 cylinders 10518 cm^3 max. 324 kW / 1900 min^{-1}	Oxi-cat SCR	Euro V
Passenger car (Audi A4, diesel)	4 cylinders 1968 cm^3 max. 105 kW / 4200 min^{-1}	Oxi-cat Diesel Particle Filter	Euro V
Passenger car (Saab 95 Biopower)	4 cylinders 1984 cm^3 max. 110 kW / 5100 min^{-1}	3-ways catalyst	Euro IV
Off-road Engine (John Deere 6068)	6 cylinders 6788 cm^3 124 kW	none	EU III A / Tier III

For heavy duty vehicles the actual test cycles are the European stationary cycle (ESC) and the European transient cycle (ETC) also know as FIGE (195). These cycles were applied for the MAN truck emission sampling. Off-road specifications are different, as a matter of fact for the characterization of the tractor engine emission, the non road transient cycle (NRTC) and non road stationary cycle (NRSC) were used. All test cycles were separately sampled. Both passenger cars (diesel and otto-cycle) were operated with the new European driving (NEDC) cycle and the common ARTEMIS driving cycle (CADC). Artemis is the acronym of the European project "assessment and reliability of transport emission models and inventory systems". The two cycles are both transient and were sampled together. A detailed description of fuels and test cycles used is provided in Table 3-2.

Table 3-2. fuels and test cycle used.

Test Vehicle	Fuel	Test Cycle
	Fossil diesel fuel (EN 590)	FIGE and ESC
	B100 (RME EN 14214)	FIGE and ESC
Heavy duty	Vegetable oil (PÖL)(V 51605)	FIGE and ESC
Euro 5	B7 (fossil diesl with 7% FAME)	FIGE and ESC
	B10 (fossil diesel with 10% FAME)	FIGE and ESC
	B7+3 (fossil diesel with 7% FAME and 3% vegetable oil	FIGE and ESC
	Fossil diesel fuel (EN 590)	NRTC and ESC
	B100 (RME EN 14214)	NRTC and ESC
Tractor engine	Vegetable oil (PÖL)(V 51605)	NRTC and ESC
EU III A	B7 (fossil diesel with 7% FAME)	NRTC and ESC
	B10 (fossil diesel with 10% FAME)	NRTC and ESC
	B7+3 (fossil diesel with 7% FAME and 3% PÖL)	NRTC and ESC
	Fossil diesel fuel (EN 590)	NEDC + CADC
Passenger car	B100 (RME EN 14214)	NEDC + CADC
Euro 5,	B7 (fossil diesel with 7% FAME)	NEDC + CADC
diesel	B10 (fossil diesel with 10% FAME)	NEDC + CADC
	B7+3 (fossil diesel with 7% FAME and 3% PÖL)	NEDC + CADC
	Fossil gasoline (DIN 228)	NEDC + CADC
Passenger car	E85 (fossil gasoline with 85% ethanol, DIN 51625)	NEDC + CADC
Euro 4	E75 (fossil gasoline with 75% ethanol DIN 15376)	NEDC + CADC
otto-cycle	E10 (fossil gasoline with 85% ethanol DIN 15376)	NEDC + CADC
	E05 (fossil gasoline with 5% ethanol DIN 15376)	NEDC + CADC
	E03 (fossil gasoline with 3% ethanol DIN 15376)	NEDC + CADC

3.4.2 Sampling at MAN

A heavy duty engine (MAN Bus engine, Euro IV, 10600 cm^3, max. 300 kW) was used for real diesel soot sampling in three different campaigns. Focus was the specific evaluation of the DPF effects. Exhaust gas was sampled by means of a microtunnel (Microtroll 4, NOVA, Germany). With this system it was possible to take a defined part of the exhaust gas and to dilute it with a known amount of fresh air. Dilution ratios were comprised between 1:4 and 1:6. Exactly 235 g h^{-1} (4 L min^{-1}) of diluted aerosol were directed to a quartz fiber filter for particle collection. As discussed in previous sections, to avoid artifact NPAH formation due to nitration of PAH on the sampling filter, two 3-channel annular denuders were positioned in front of the filter. The denuder system capacity was the limiting factor, therefore 4 L min^{-1} were the highest flow allowed. Denuders were regenerated every 30 or 60 minutes depending on the NO_2 concentration which after dilution never exceeded 30 ppm. A scheme of the sampling system is illustrated in Figure 3-9.

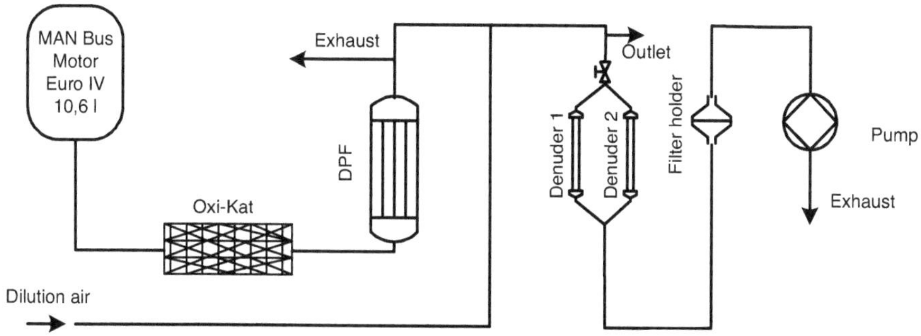

Figure 3-9. Scheme of the sampling system at the MAN engine test bench.

The motor was operated at the points 9 and 10 of the ESC cycle. Point 9 (B25) simulates a load of 25% at 50% of engine speed. With the motor used for this sampling, 50% of engine speed corresponds to approx. 1500 rpm. Point 10 (C100) simulates a load of 100% at the maximum engine speed (approx. 2000 rpm). All 13 ESC points are illustrated in Figure 3-10

The aftertreatment system was an oxidation catalyst coupled to a DPF. The DPF (AC 200 cpsi, Duratrap) was a wall-flow filter with a cordierite based honeycomb structure. The oxidation catalyst mounted in front of the DPF oxidized the NO to NO_2 with an efficiency comprised from 20% and 60% depending on the temperature.

The three microtunnel sampling points were three:

A Raw (before oxi-cat)

B Before DPF (after oxi-cat, before DPF)

C After DPF

Sampling time was 15 min for sampling points A and 30 min for points B and C. DPF filtration efficiency was always comprised between 90% and 99%. Sampled soot mass was comprised between 0.06 mg and 0.5 mg depending on sampling positions. Six quartz fiber filters for each microtunnel sampling point (three for B25 and three for C100) were loaded for PAH and NPAH analysis.

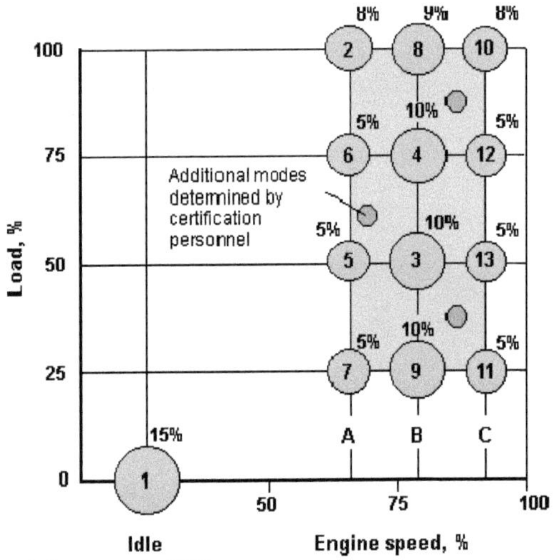

Figure 3-10. European stationary cycle (ESC).

3.5 Analysis

3.5.1 Reagents, solvents and standards

Aluminum oxide with 5% platinum was supplied by VWR International (Darmstadt, Germany). Dichloromethane (DCM), cyclohexane, methanol (MeOH) and toluene (all of analytical grade) were obtained from Fluka (Buchs, Switzerland). Water was taken from a high-purity water system (MilliQ plus 185, Millipore, Eschborn, Germany). Standard solutions of 1,3-dinitropyrene, 7-nitrobenzo(a)anthracene, 6-nitrobenzo(a)pyrene, 1-nitronaphthalene, 2-nitronaphthalene, 3-nitrofluoranthene, 3-nitrophenanthrene, (10 mg/L of each in cyclohexane) and 1,6-dinitropyrene (10 mg/L in toluene) were purchased from Ehrenstorfer (Augsburg, Germany). PYR, 2,7-dinitrofluorene (2,7-DNFLU), 1,5-dinitronaphthalene, 1-NPYR (solid material, 99% purity) were received from Sigma-Aldrich (Deisenhofen, Germany). 1-nitrobenzo(a)pyrene and 3-nitrobenzo(a)pyrene were obtained by synthesis (132,196). To prepare standard solutions from these solid compounds the necessary amount of the neat material to compensate the difference of the purity of the compound to 100% has been calculated. Single components were weighed with an electronic microbalance (Mettler AT261, Giessen, Germany) placed into a volumetric flask and filled to the mark with cyclohexane (Roth, Germany) resulting in stock solutions of 10 mg/L. Mixed working standards were prepared from these single-compound solutions.

3.5.2 Sample extraction and cleaning

Each filter was at first spiked with internal standard (I.S.). Amount and type of the internal standard were dependent on the kind of sample to analyze. In case of real diesel soot samples 2,7-DNFLU was preferred. 2,7-DNFLU is a well detectable NPAH rarely present in real samples or present in traces that can be neglected in comparison to the amount that is spiked on the sample. For laboratory samples, also 3-nitrophenanthrene and 7-nitrobenzo(a)anthracene were used. The reason was that those compounds showed the best similarity in recovery tests with 1-NPYR and 6-NBaP, the reaction products of major interest.

After evaporation of the I.S. containing solvent, the filters were transferred into to a 25-mL pear-shaped distilling flask and extracted in a 20-mL DCM/MEOH/toluene (1/1/1; v/v/v) mixture for 30 minutes under ultrasonic agitation. The filter were removed and the extract evaporated in a rotary evaporator under vacuum (50°C, 80 hPa) to a volume of about 0.5 mL. The residual solution was transferred to a glass column (length 80 mm, diameter 11 mm) filled with 1 g silica gel 60 (Merk, Darmstadt, Germany) that was preconditioned with 8 mL of cyclohexane. To ensure quantitative transfer of all the analyte, the distilling flask was washed with 0.5 mL of cyclohexane. The washing solution was passed through the silica column. The analyte elution was carried out with 4 mL cyclohexane and 12 mL cyclohexane/DCM (1:1; v/v). The eluate was evaporated in a rotary evaporator under vacuum (50 °C, 80 hPa) to 0.5 mL, quantitatively transferred to an HPLC vial and evaporated further to a volume of approx. 10 µl under gentle N_2 gas flow. Before analysis, 490 µl of cyclohexane were added to the sample containing vial.

3.5.3 PAH analysis

PAH analysis was performed by means of HPLC with fluorescence detection. PAH are strong fluorescent compounds and can be easily detect by setting the detector at the proper excitation emission wavelength. Mobile phase was acetonitrile/water. The mobile phase gradient is illustrated in Table 3-3

Table 3-3. Solvent gradient for PAH analysis.

Time (min)	Solvent A: Water [%]	Solvent B: Acetonitrile [%]
0	40	60
10	0	100
35	35	100
40	STOP	STOP

Table 3-4.Excitation and emission wavelengths for PAH fluorescence detection, retention times (RT) and limit of detection.

PAH	λ_{ex} [nm]	λ_{em} [nm]	RT [min]	LOD [µg/l]
Naphthalene	275	322	5.83	0.30
Acetnaphthene	275	322	8.55	0.26
Fluorene	275	322	8.90	0.09
Phenanthrene	275	322	9.97	0.69
Anthracene	252	370	11.10	0.19
Fluoranthene	277	462	12.17	0.12
Pyrene	277	462	12.99	0.06
Triphenylene	270	380	14.03	0.07
Benzo[a]anthracene	270	380	15.51	0.06
Chrysene	270	380	16.17	0.02
Benzo[k]fluoranthene	297	410	18.08	0.12
Benzo[b]fluoranthene	297	410	19.15	0.07
Benzo(a)pyrene	297	410	20.23	0.05
Dibenz[a,h]anthracene	297	410	22.06	0.26
Benzo[g,h,i]perylene	297	410	23.36	0.16

3.5.4 Nitro-PAH analysis

NPAH analysis was carried out by means of HPLC coupled to a fluorescence detector. The weak fluorescence exhibited by NPAHs requires a reduction of the nitro-PAHs to the corresponding fluorescing amino-PAHs (112). Thus, a reduction column (Macherey-Nagel, 50 mm x 4 mm) was densely packed with 5% Pt on Al_2O_3 and placed after the reverse-phase analytical column (Restek pinnacle PAH II 150 mm x 3 mm, 4 µm) used for the separation. The reduction column was dry-filled. It needed to be activated by 20 – 30 injections before showing a constant activity. Reduction column regeneration was requested as soon as the peak form and/or peak intensity got worse/reduced. Heating overnight at 500°C was often a solution, nevertheless sometimes was it necessary to fresh re-pack the column. The separation column temperature was kept constant at 25°C, while the reduction column was at 90°C, the optimum temperature (197).

Mobile phase was methanol/water, with a starting mixture ratio 60/40 (v/v). Many solvent gradient, for the application to different samples, have been tested and used. As example, the solvent gradient which was most frequently used is described in Table 3-5. It allows the separation of 12 NPAH within 50 min with a flow of 0.7 mL min^{-1}. The resulting chromatogram is illustrated in Figure 3-11. A slight modified solvent gradient was used to allow simultaneous separation of 12 NPAH and two PAH (PYR and BaP). A chromatogram is illustrated in Figure 3-12. Standard solutions for calibration were prepared by dilution of the mixed standard stock solutions (10 µg mL^{-1} of each NPAH in cyclohexane) with cyclohexane. Standard solutions with NPAH concentration of 0.1, 0.5, 1.0, 5.0 and 10 µg/l were injected three times each. Injection volume was 20 µl. Average and standard deviation were calculated and the best fit curves were interpolated. Samples were analyzed in batches of 10 - 15 vials. The system was calibrated before every batch was started. The concentration detection limits were determined accordind IUPAC. The mass detection limits follow from multiplication of the concentration detection limits with the HPLC injection volume (20 µl). Retention times and limits of detection are shown in

Table 3-6.

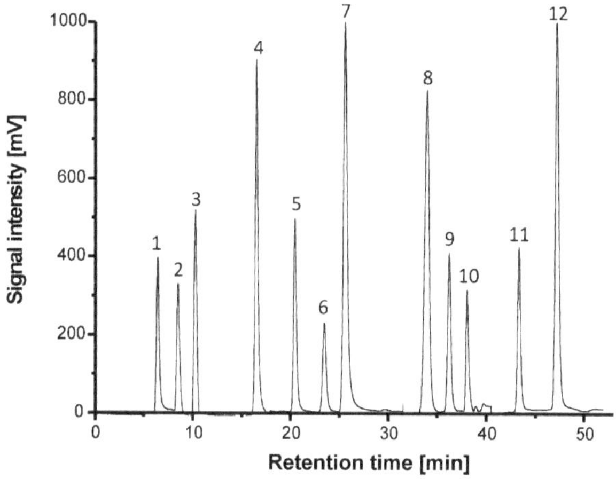

Figure 3-11. Chromatogram of a NPAH standard, with the solvent gradient method of table 3-1. Peaks table at page 129

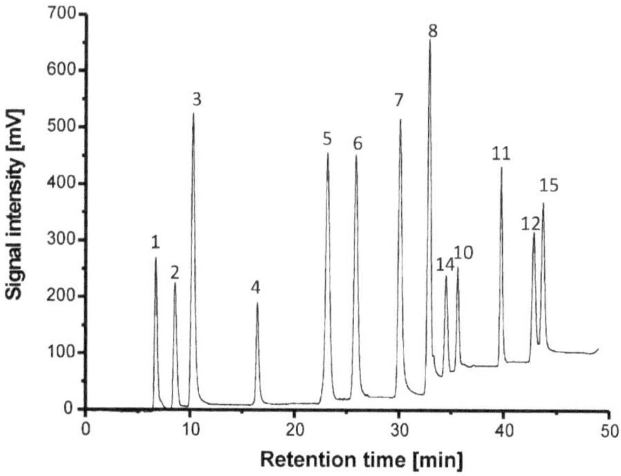

Figure 3-12. Chromatogram of a NPAH standard, with the modified solvent/gradient method. Peaks table at page 126.

Table 3-5. Solvent time program for NPAH analysis.

Time (min)	Solvent A: Water [%]	Solvent B: Methanol [%]
0	40	60
5	40	60
15	35	65
25	35	65
40	0	100
46	0	100
48	40	60
52	STOP	STOP

Table 3-6. Excitation and emission wavelengths for NPAH fluorescence detection, retention times (RT) and limits of detection.

Nitro-PAH	λ_{ex} [nm]	λ_{em} [nm]	RT [min]	LOD [µg/l]	LOD [pg]
1,5-DNNAP	231	290	6.533	0.093	3.5
1-NNAP	243	429	8.969	0.072	3.6
2-NNAP	234	403	10.21	0.078	1.6
2,7-DNFLU	292	387	16.81	0.071	1.9
2-NFLU	290	365	23.62	0.094	0.21
3-NPHE	254	440	25.77	0.096	0.90
1,6-DNPYR	369	442	30.19	0.078	3.0
1-NPYR	360	430	32.72	0.066	0.12
3-NFLT	300	530	34.37	0.076	4.0
7-NB[a]A	300	475	36.06	0.062	2.7
6-NB[a]P	420	475	39.71	0.051	0.4
1-NB[a]P	420	475	42.65	0.076	1.52
3-NB[a]P	420	475	42.65	0.076	1.52

4 Results and discussion

4.1 Aerosol and gas mixture characterization

Pre requisite of this study was the production of an artificial aerosol by means of a spark-discharge soot generator and its coating with a sub-monolayer of PYR or BaP. The aim was to dispose of a proxy for diesel soot particles for nitration experiments under simulated diesel exhaust conditions. The constancy of the system and therefore the reproducibility of the produced aerosol were of central importance for the execution of proper nitration experiments. The production of gas mixtures containing NO_2 in known concentration was an issue and the sampling without the formation of artifacts had to be also guaranteed. Therefore two orthogonal methods for the NO_2 concentration analysis were used and the performances of different denuding systems tested.

This section summarizes the most important results regarding the aerosol and gas mixtures characterization in term of size distribution, PAH coating, NO_2 concentration and denuding systems.

4.1.1 Aerosol size distribution

A SMPS system was constantly assisting the aerosol production to control important parameters like size distribution, total number of particles and therefore the total surface area. By the experiments with PYR-coated soot, the aerosol exhibited a log-normal particle size distribution with a mobility equivalent count median diameter of 90 nm, a mode of 84 nm and a geometric standard deviation of 1.6, in agreement with previous findings (41). In the succesive BaP-coated soot experiments, particles were somewhat bigger with a mobility equivalent count median of 107 nm, a mode of 101 nm and a geometric standard deviation of 1.6. This increase is not addressed to the BaP coating, rather to the higher spark discharge frequency applied in the BaP series in comparison to the PYR ones. Actually, differences in the particle size distribution between PAH coated and uncoated soot could not be noticed with neither PYR nor with BaP.

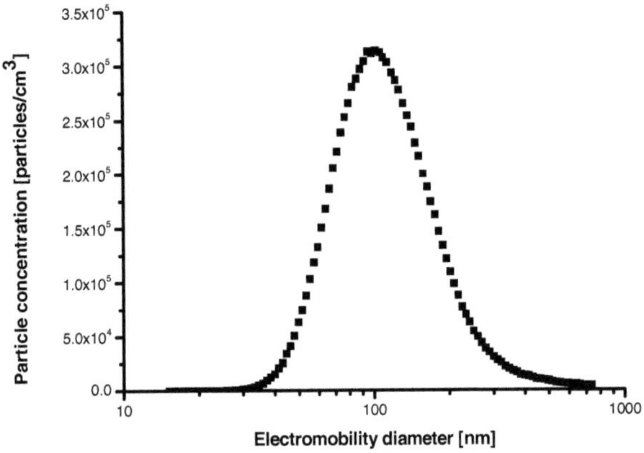

Figure 4-1. Typical size distribution of the BaP-coated spark discharge soot aerosol used in this study.

During the three years of this work, more than 2000 SMPS scans have been collected to check the stability of the system. For example, in Figure 4-1 a typical size distribution of BaP-coated GfG soot aerosol is shown.

4.1.2 Estimation of PAH surface coverage

The amount of PYR and BaP coating was separately measured at four PAH source temperatures ranging from 20°C to 60°C, and from 60°C and 100°C for PYR and BaP respectively. A typical exponential dependence PAH amount on soot is observed as a function of temperature i.e. PAH vapor pressure. The vapor pressure of pyrene is higher than that of benzo(a)pyrene. Because of that, PYR coating experiments were made in a temperature range lower than that for BaP. Figure 4-2 and Figure 4-3 show these results.

Systematic filter loading for PYR nitration experiments was made with a coating temperature of 20°C. This temperature assured a good PYR coating (2.18 ± 0.45 μg/mg). Systematic filter loading for BaP nitration experiments was instead made maintaining the BaP reservoir at 100°C. This temperature was used by necessity chosen to get enough BaP adsorbed on particles.

Nevertheless BaP coated mass (0.40 ± 0.14 µg/mg) was about 5 times lower than PYR. Reducing the soot amount to get an higher BaP/soot ratio was not practicable. Actually the deposited soot amount of ca. 0.5 mg in 1 h (see 3.1.2) was considered as the lowest mass that could be easily handled on quartz fiber filter.

The main problem associated with coating at 100°C is that BaP, due to the exponential increase of the vapor pressure, is in a range where a slight difference in temperature produces a large effect in the vapor pressure. Thus, the difficult to perfectly control the BaP source temperature and the isolation between the BaP reservoir and the ring mixing gap, (see Figure 3-3. PAH aerosol generator system) resulted in remarkable differences in the sublimation of BaP. The coating was therefore affected from a relative standard deviation of 35%.

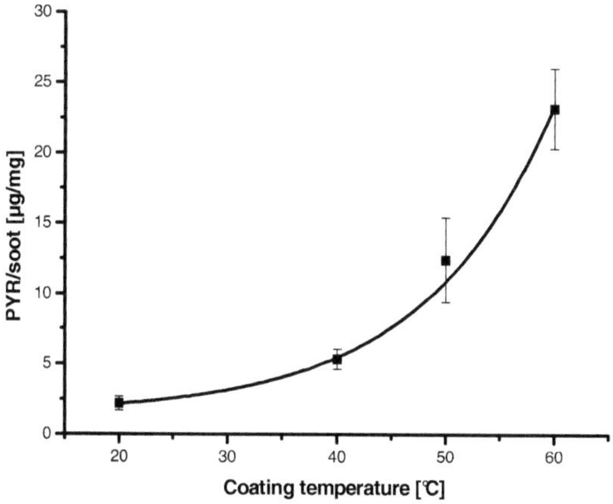

Figure 4-2. Dependence of pyrene/soot coating ratio on PYR source temperature. Data and error bars represent the means and standard deviation of four to five repeated measurements. The exponential fit is based on the Clausius-Clapeyron equation.

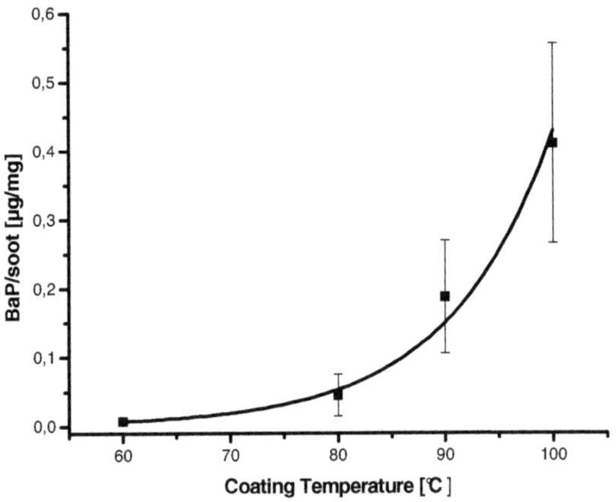

Figure 4-3. Dependence of BaP/soot coating ratio on BaP source temperature. Data and error bars represent the means and standard deviation of three to six repeated measurements. The exponential fit is based on the Clausius-Clapeyron equation.

The PYR and BaP surface coverage in unit of monolayers has been calculated. That was necessary for a more complete characterization of the aerosol. Moreover it has to be made sure that the coated PAH was adsorbed as sub-monolayer. Actually, such a surface coating is similar to that observed with real diesel soot, where the amount of adsorbed PAHs is hardly enough to form a complete monolayer.

The estimation for PYR-coated soot aerosol was obtained following the method described by Pöschl *et al.* (41). The mean particle number concentration in the sampling system was 2×10^{7} cm^{-3}. To determine the average particle mass m_p, it was assumed that the number of particles collected on the filter equals the average number concentration in the sampling flow multiplied by the sampling volume of 573 L (1 h at 9.6 L min^{-1}). That led to a total number of collected particles of 1.15×10^{13}. By dividing the total sampled mass by this value, an average estimate for m_p of 7.4×10^{-17} g was calculated. An estimate for the average particle surface area of the GfG soot aerosol s_p, has been obtained by multiplying m_p by the literature value of the GfG soot used in this study (395 m^2 g^{-1}) (174), which led to an s_p of 2.9×10^{-10} cm^2. To calculate the PYR surface concentration of GfG soot particles $[PYR]_s$, the total amount of PYR molecules was divided by the total number of particles deposited on the filter and by the average particle surface area. At a temperature of 20°C, this leads to a value for $[PYR]_{s20°C}$ of 1.65 x 10^{12} cm^{-2}. The molecular surface area of one PYR molecule is reported to be 1.91 x 10^{-14} cm^2 (198). Assuming an even distribution of the PYR molecules, one PYR surface monolayer could be obtained with a $[PYR]_s$ of 5.2 x 10^{13}. Dividing $[PYR]_{s20°C}$ by this value, results a surface coverage of 0.032 monolayer. To confirm this value, the surface coverage has been also calculated with an other method. In fact, since the aerosol size-distribution and the mean particle number concentration in the sampling system are known, the mean aerosol surface area s_a, was calculated to be 5.2 x 10^{-3} cm^2. Under the same assumptions valid for the calculation of m_p, results an estimate of the total aerosol surface area s_{tot} of 2.98 x 10^3. By multiplying the molecular PYR surface area (1.91 x 10^{-14} cm^2) by for the mean number of PYR molecules, one can obtain the total PYR surface area. At 20 °C it is 1.06 x 10^{-2} cm^2. Dividing this value by s_{tot}, an estimation of a surface coverage in units of monolayers of 0.035 can be obtained. This value is in good agreement with the value calculated above.

An estimation of the BaP surface coverage was obtained in the same way as for PYR. The mean particle number concentration in the sampling system was 1 x 10^7 cm^{-3}, therefore s_p was 3.4 x 10^{-10} cm^2. Considering the BaP molecular area = 2.31 x 10^{-14} cm^2 (198), at 100°C the $[BaP]_s$ was 3.18 x 10^{11} which results in 0.0075 monolayers.

4.1.3 Determination of the NO_2 concentration

Known amounts of NO_2 in the range of 0.1 ppm – 6 ppm, were generated by mean of two permeation tubes. Permeation rates were determined by weighing the tubes at different time intervals and calculating the mass of nitrogen dioxide that had permeated (Figure 4-4 and

Figure 4-5). For the purpose of converting a permeation rate into a concentration, two possibilities were given. The first was to convert the permeated NO_2 mass into a volume using the ideal gas law and to calculate the ppmv (v/v) after the dilution with a know volume of the gas mixture used (95% N_2 and 5% O_2). The second was to simply divide the mass of permeated NO_2 by the density of the gas mixture used for the dilution, which at 20°C is 1.16 g L^{-1}. In this way a concentration in ppm (m/m) is obtained, which means a mass ratio (µg of NO_2 per g of gas mixture).

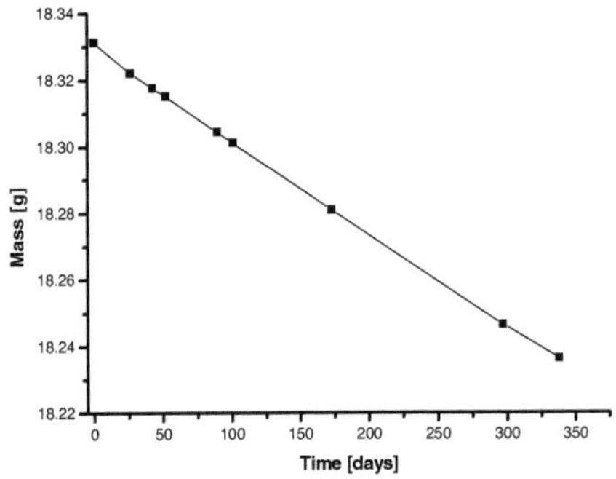

Figure 4-4. Mass loss over one year (permeation tube A).

Table 4-1. Permeation rates and ppm of NO_2 calculated in 1 L of N_2/O_2 gas mixture at 20°C (Tube A).

Day	Rate [µl NO_2 min^{-1}]	ppm NO_2
0	-	-
26	0.126	0.208
42	0.099	0.163
52.	0.096	0.159
89	0.103	0.170
100	0.106	0.175
172	0.102	0.168
296	0.101	0.166
337	0.100	0.166

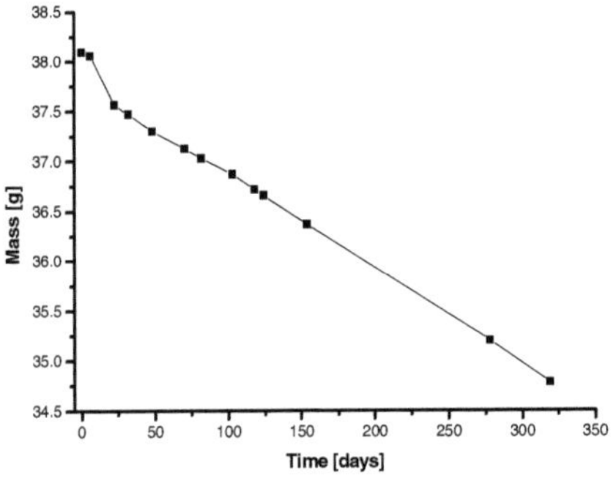

Figure 4-5. Mass loss over one year (permeation tube B).

Table 4-2. Permeation rates and ppm of NO_2 calculated in 1 L of N_2/O_2 gas mixture at 20°C (Tube B).

Day	Rate [µl NO_2 min^{-1}]	ppm NO_2
0	-	
6.00	2.48	4.09
22.0	11.2	18.4
31.7	3.52	5.81
47.9	3.84	6.34
69.9	2.89	4.77
81.3	3.02	4.98
102.5	2.72	4.49
117.5	2.69	4.44
123.9	2.81	4.64
153.7	2.96	4.88
277.7	3.05	5.04
318.8	2.66	4.38

All the concentrations in this work have to be intended as ppm (m/m). By dividing the ppm value by the ppmv value, a conversion factor of 1.65 is obtained, a useful tool to easily convert ppm of NO_2 to ppmv and vice versa. Permeation rate from tube A, was more constant than permeation tube B. The reason could be its thicker PTFE membrane, making it less susceptible to variation of temperatures or atmospheric pressure. Either way a satisfactory consistency in the permeation rate could not be achieved before 3 - 4 weeks after re-filling. The so calculated NO_2 concentrations $[NO_2^P]$ had to be confirmed by an orthogonal method. For this purpose the $[NO_2]$ has been analyzed also by mean of the wet analytical method already described in section 3.2.2. In Figure 4-6 is the calibration curve used for one of the analysis that were periodically carried out. This was especially necessary a few weeks after re-filling the tubes or before a long nitration experiment series. The blank corrected absorbance at 550 nm was plotted against the mL of $NaNO_2$ standard solution used for calibration. Beer´s law is followed and since 1 mL of standard solution is equivalent to 4 µl NO_2 (184), a volume of NO_2 could be calculated.

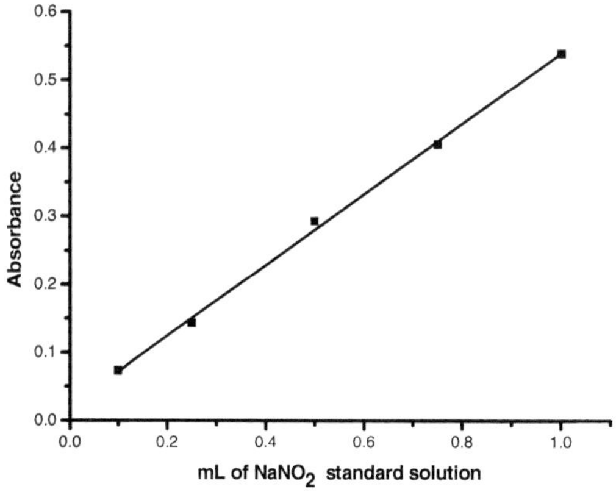

Figure 4-6. Calibration curve for the wet-chemistry analysis of $[NO_2]$ by Saltzman method.

The obtained value in ppmv can be easily converted in ppm of NO_2^{wet} by mean of the conversion factor 1.65, and directly compared with the NO_2 permeation rate also expressed in ppm. [NO_2^{wet}] was always slightly lower as [NO_2^{P}]. However, the difference did never exceeded 10% over 5 measurements, and was generally lower than 5%. The reason could be that not all the mass that permeated the tubes membrane was NO_2, the only specie that could be detected by the wet-chemical method. Traces of NO could be present or HNO_3 or HNO_2 could be formed during the re-filling operations, when gaseous NO_2 become condensed with liquid nitrogen in the tubes. Condensed water, simply coming from the surrounding ambient air could be present too, thus reacting with NO_2.

4.1.4 NO_2 scrubbing by annular denuders

NO_2 removal from the sampling line was a central issue both for real diesel soot sampling and for laboratory experiments. The chosen method was to scrub NO_2 by mean of multi-channel annular denuders coated with KI or guiacol. They had to guarantee a high NO_2 removal efficiency under sampling conditions, so the performances of such systems had to be tested and the effect of parameters like [NO_2] and flow velocity on the scrubbing efficiency had to be verified. Generally high efficiencies could be achieved by reducing the flow rate, thus increasing the residence time in the denuder annulus. In this way NO_2 molecules have more time to diffuse to the coated wall, becoming removed from the gas flow. On the other hand, by lowering the aerosol amount which is directed through the denuder, the total soot mass collected was reduced to levels too low for a proper sampling and analysis. There was then a trade-off between the reduction of the flow rate for the increase of the NO_2 removal efficiency and the amount of soot that can be collected. An acceptable compromise was needed.

The approach used for the efficiency tests was to measure the [NO_2] by means of the Saltzman method or of the NO_x analyzer (see section 3.2.3) before and after the denuder to obtain NO_2 scrubbing efficiency curves. The denuders tested were all three-channel and previously described in section 3.2.4

First experiments involved one denuder, at $[NO_2]$ = 3 ppm and a flow of 0.9 L $^{-1}$. The two measurements are shown in Figure 4-7. The NO_2 removal efficiencies after 120 min were still higher than 85%.

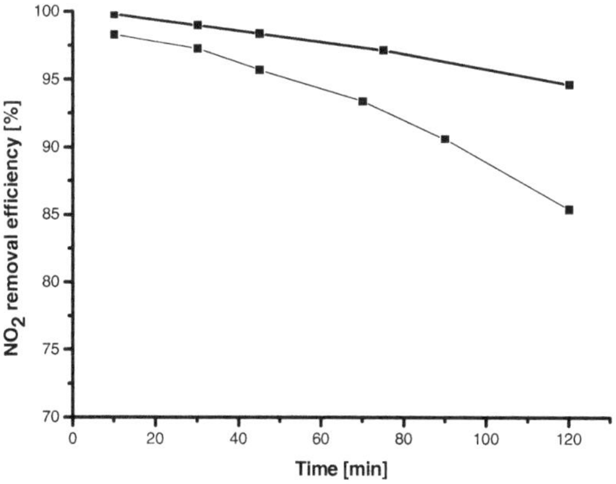

Figure 4-7. NO_2 removal efficiency of one denuder tested at $[NO_2]$= 3 ppm and at a flow of 0.9 L min^{-1}.

To increase the capacity of the denuding system and to allow higher concentrations and flow rates, two denuders in parallel have been used. The concentration of interest at which nitration experiments in the DPF simulation were carried out was 120 ppm. It was necessary to guarantee a high NO_2 removal for at least 1 hour, the actual duration of the experiments and simultaneously be able to collect enough soot (> 0.10 mg). For brevity, only some of the results will be presented.

Figure 4-8 shows the results of two tests with two disposed in parallel denuders and 120 ppm of NO_2 at a total flow of 5 L min^{-1}. Under these conditions, the calculation of the Reynolds number demonstrated that the flow in the tubes was still laminar, but the $[NO_2]$ was too high to be efficiently reduced at that flow rate. At the beginning of the test, NO_2 removal efficiency was only 90% and within 30 min dropped to 50% letting 60 ppm of NO_2 passing through.

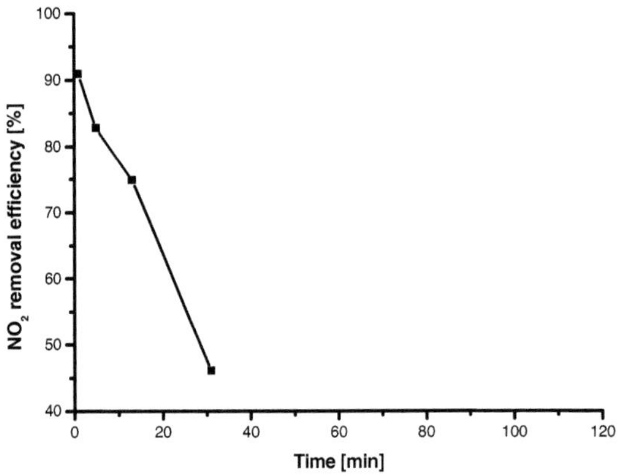

Figure 4-8. NO_2 removal efficiency of two parallel three-channel denuders at total flow of 5 L min^{-1} and $[NO_2]$ = 120 ppm and.

After several trial and error tests, a flow rate of 1.5 L min^{-1} has been chosen as best compromise between an acceptable amount of sampled soot and a good NO_2 removal efficiency (Figure 4-9). Under these conditions, four independent tests have been carried out. Two of them were measured with the Saltzman method and the other two with the NO_x-analyzer. Efficiencies after 60 min usage were comprised between 77% and 95%.

The differences have to be addressed to the KI coating, which was not uniform between the different denuders of the four tests. A partial solution to this problem was the multi-coating. It consisted of coating twice or three times to obtain a thicker coating layer. As matter of fact, a slightly increase of the NO_2 removal capacity but in particular an increase of the uniformity in the efficiency in different tests could be obtained. In this way, average efficiencies of ca. 89% ± 6% after 1 h usage, could be achieved.

A saturated solution of guiacol was also tested. Guiacol was much more effective than KI at NO2 concentrations of few ppm, which was removed with more than 95% efficiency for more than two hours. On the other hand, the coating quality was poor and it was not so effective at NO2 concentrations > 50 ppm. Since during nitration experiments guiacol coating has never been used, results are here not reported in detail. Briefly, it can be summarized that guiacol has an higher efficiency than KI but a lower capacity. As a consequence, at high NO2 concentration it removes NO2 efficiently just for few minutes and then the efficiency decreases rapidly. Moreover guiacol`s interaction with PAH could not be tested and it has a negative impact on the stability of the denuders (it corrodes the adhesive material which glued the glass tubes together). For all these reasons KI has been preferred.

Regarding sampling of real diesel soot from an engine test-bench like that at MAN, different operating conditions had to be taken in account. Sampling was done by pulling a sample from a microtunnel system (see section 3.4.2) where the exhaust gas was already diluted with a dilution factor depending on the driving cycle. Because of that, [NO_2] in the sampling line was sensibly lower than the tailpipe-out [NO_2] never exceeding 30 ppm, and in most of the cases was much lower. Therefore the flow rate through the denuder could be higher, thus allowing the collection of a sufficient mass of soot. Considering the lower NO_2 concentration and driving cycles of 30 min instead of 1 h, sampling was done with a total flow of 4 L min^{-1}.

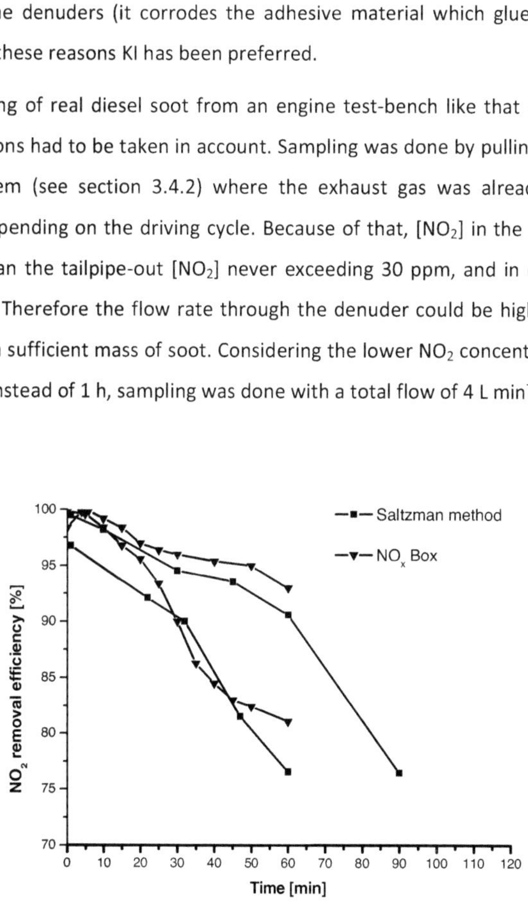

Figure 4-9. NO_2 removal efficiency of two parallel three-channel denuders I at total flow of 1.5 L min^{-1} and [NO_2] = 120 ppm.

4.2 Nitration experiments

Core of the experimental work carried out in this study, was the investigation of the reaction between soot-adsorbed PAHs and NO_2, with a focus on the reaction products, especially NPAHs. A part of the experiments was conceived to investigate the artifacts formation of NPAH under sampling conditions. Others were designed to understand whether or not a post-combustion formation of NPAH can take place within a DPF structure.

At the beginning the approach was to first coat GfG soot with PYR or BaP and to collect it on a quartz fiber filter. After this step, the loaded filters were exposed to a NO_2 containing atmosphere. The proceeding of the heterogeneous reaction was measured by stopping the reaction at given times and performing HPLC analysis of the extracts to get the necessary information. Effect of reaction time, NO_2 concentration and reaction temperature have been investigated.

A second and successive approach was to simulate both real DPFs systems with a deep-bed filtration (PM-Kat®) mechanism and with a wall-flow one (cordierite honeycomb filter). The PAH coated soot was directed to the DPFs, and the outlet aerosol was collected and analyzed. The inlet aerosol being NPAH-free, finding NPAH after the DPF would have meant a NPAH formation within the DPF structures. Focus was on reaction temperature and on nitration products of BaP.

4.2.1 Nitration reaction kinetics: Pyrene-coated soot

The influence of the reaction time on the nitration reaction in PYR-coated soot was studied at 20°C in the presence of 4 different NO_2 concentrations: 0.11 ppm, 1 ppm, 2 ppm and 4 ppm. Focus of this experiment was the artifact NPAH formation under sampling conditions since collected PAHs can be nitrated on the sampling filter by the NO_2 constantly passing through it. Moreover, the investigation of the heterogeneous interaction between NO_2 and PYR on soot is of interest for diesel exhaust at low temperature (e.g. cold start).

NO_2 concentrations have been chosen to cover NO_2 amounts that range from levels representative of a polluted city atmosphere (0.11 ppm, EU limit 0.20 ppm), to levels hat are considered as the lower limit of the engine-out NO_2 concentration (4 ppm). Values in between or slightly above can be typically found in diluted diesel exhaust sampling. Reaction times have been chosen taking in considerations typical sampling times (30 min, 60 min). Higher reaction times are representative of particles trapped in the exhaust aftertreatment (e.g. DPFs between two regenerations), where they have a longer exposure time to NO_2.

Under these conditions formation of 1-NPYR is evident and it is the only NPAH detected with the current analysis method and in agreement with previous findings (199-201). In Figure 4-10 the amount of produced 1-NPYR, normalized to the soot amount deposited on the filter, is plotted versus reaction time. The data points represent three to five repeated measurements performed at 30, 60, 120, and 180 minutes. The data at x = 0 minutes represent the blank values. No contamination was evident. The amount of 1-NPYR formed is linearly correlated with the reaction time. The 1-NPYR formation rate v is determined from the slope of the linear least-square fit like the one shown in Figure 4-10 (see Table 1).

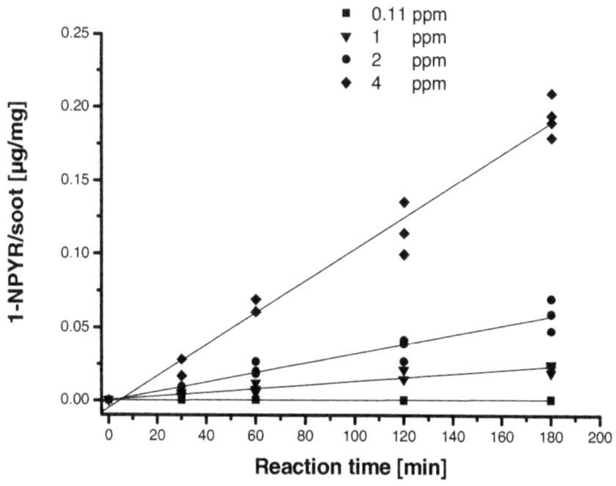

Figure 4-10. Normalized amount of 1-NPYR on soot aerosol versus the reaction time. The straight lines are linear least-square fits based on pseudo-zero order rate equations. Pyrene amount was 1.85 ± 0.37 µg for each mg of soot.

Table 4-1. 1-NPYR formation rate v obtained from the curve fits displayed in Figure 4-10.

$[NO_2]$ (ppm)	v ($pg_{1\text{-}NPYR}\, mg_{soot}^{-1}\, min^{-1}$)
0.11	4.20×10^{-3}
1	1.23×10^{-1}
2	3.23×10^{-1}
4	1.08

In Figure 4-11 the pseudo-zero order reaction rate v of the four kinetic measurements performed is plotted versus the NO_2 concentration. The error bars represent the standard error of the linear least-squares fit, which is found to range between 3% and 8%. The rates increase more than linear with $[NO_2]$, which is indicative of the underlying reaction mechanism being not a simple one.

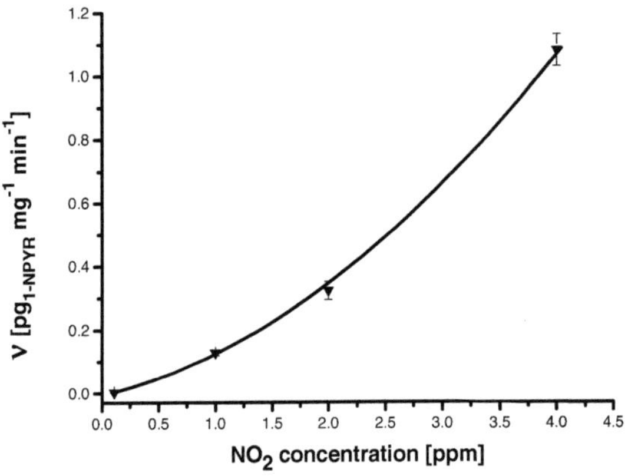

Figure 4-11. 1-NPYR formation rates as a function of NO_2 concentration. The error bars represent the standard error of the linear least-squares fit in Figure 4-10, the curve is a nonlinear least-square fits.

Previous studies estimated the decay corresponding to the reaction of adsorbed PAHs on diesel particles with NO_2 versus reaction time (36). In that case, the observed plateaus in the graphs exhibiting degradation kinetics were attributed to the complex diesel matrix causing a non-availability of the adsorbed PAHs and a minor accessibility to the NO_2. In our case we are assuming the PYR molecules are more accessibly arranged on the particle surface and, at the highest reaction time of 180 minutes and at the highest NO_2 concentration of 4 ppm ca. 10% of PYR is transformed in 1-NPYR. Thus, it is in good agreement with the linearity observed.

Kristovich *et al.* (202) investigated the nitration reaction of BaP with NO_2 on several substrates like silica particles, zeolites and coal fly ash. They found also that the nitration reaction with NO_2 did not occur on silica particles, in contrast with other findings of Wang *et al.* (199,203). They addressed the difference to the possibility to form HNO_3 which was considered of central importance in the nitration reaction. Because Kristovich used dried silica particles, HNO_3 could not be formed. On the contrary, NPAH were formed on coal fly ash and thermal-treated zeolites. They found that the production of nitroderivates increased with time and proposed a mechanism where BaP-radical-cations, formed by the interaction of BaP with active sites (broken alumina-oxygen bonds on the substrate), reacted with NO_2 with formation of HONO and mono-nitroderivates of BaP. This mechanism could be suitable for our conditions only if such active sites would be present on GfG soot at room temperature. Moreover, formation of 1-NPYR would be limited by the number of active sites where the PYR-radical cation could be formed rather than by the $[NO_2]$. Therefore the dependence on the $[NO_2]$ found in our study could be hardly explained.

Since our nitration experiments occurred in presence of H_2O, NO_2 could react heterogeneously on the soot particle with it to form HNO_3 (204). The subsequent dissociation would provide H^+ which could react again with NO_2 to form the electrophilic species HNO_2^+. NO_2, due to its equilibrated dimerization with dinitrogen tetraoxide (N_2O_4) (205), forms also $HN_2O_4^+$. However it would react in the same way as HNO_2^+, which for simplicity will be the only species considered. After formation, HNO_2^+ would then attack PYR in position 1-, which has the highest electron density to form $PYRHNO_2^+$ intermediate. 1-NPYR is formed after elimination of HONO and H^+. This is a typical electrophilic mechanism and has been already proposed for pyrene nitration (199). Another possibility is the formation of HONO (206), which would again react with the H^+ originated from HNO_3 dissociation forming the nitronium ion (NO_2^+) and H_2O (203). NO_2^+ would then react further to form an $PYRNO_2^+$ intermediate and after elimination of H^+ obtain again 1-NPYR (207). These mechanisms are compatible with the conditions of the present study. Nevertheless such a mechanism would implicate an induction period for the formation of a sufficient amount of acid to provide enough H^+ to start the reaction. Such induction period should increase by reducing the [NO_2] like observed by Wang *et al* (199). That is hardly compatible with the formation of 1-NPYR observed, which is linear at all concentrations tested. Moreover, nitration reaction with formation of nitroderivates in presence of NO_2, occurred in absence of H_2O as well (see section 3.3.2). Therefore, it can be concluded that nitration reaction could happen following an aromatic electrophilic substitution mechanism. It is not excluded that it occurred to some extend. Either way, it can not be the only ongoing reaction mechanism and there are evidences that NO_2 could act as nitrating agent even in absence of HNO_3, attacking the aromatic ring directly.

The mechanism of reaction, which would better describe the nitration observed, is the radical addition elimination mechanism already proposed for smaller molecules as benzene, naphthalene or anthracene (208,209) and for pyrene adsorbed on silica particles (201). The proposed mechanism of reaction is presented in Figure 4-12.

Such a mechanism requires a $\bullet NO_2$ radical attack to the aromatic ring to form $\bullet NO_2PYR$ intermediate. The radical would become recombined from a second $\bullet NO_2$ and form an instable (because of the lost of aromaticity) $2NO_2PYR$. The aromatic ring would then be re-established by the expulsion of a HONO molecule to form the final NO_2-PYR.

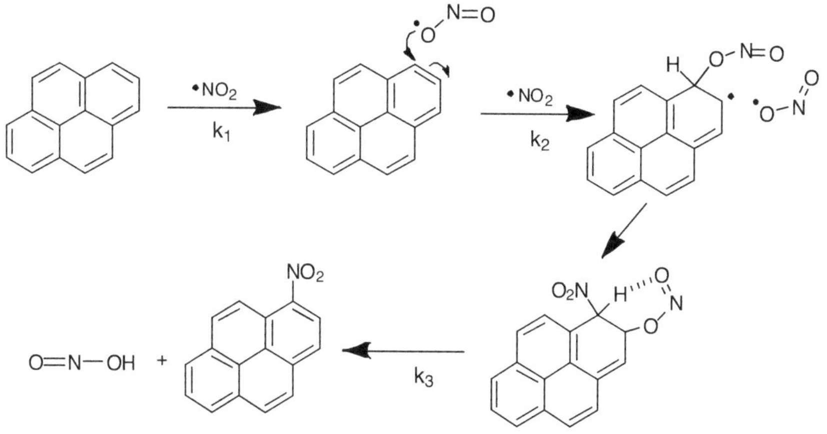

Figure 4-12. Proposed mechanism for the nitration of pyrene.

In addition to that, complex soot NO_2 interactions (e.g. first NO_2 uptake with subsequent nitration of PYR) could influence the nitration reaction (41,210). Even if further studies are necessary to confirm the nature of the reaction, some important parameters can be calculated. The 1-NPYR formation rate can be described by the following equation

$$v = k\,[NO_2]^a[PYR]^b \qquad (1)$$

where *a* and *b* are the partial reaction orders with respect to NO_2 and PYR. Assuming the presence of excess PYR, eq. (1) can be approximated as follow

$$v = k_{app}[NO_2]^a \text{ where } k_{app} = k[PYR]^b \qquad (2)$$

Converting eq. (2) to its logarithmic form, one obtains a linear equation (represented in Figure 4-13) where slope *a* is the partial reaction order with respect to NO_2 and the intercept, the Log k_{app}

$$\text{Log } v = \text{Log } k_{app} + a \text{ Log } [NO_2] \qquad (3)$$

By calculating the slope and the intercept of the linear equation in Figure 4-13, a partial order of reaction $a = 1.53$ and $k_{app} = 1.22 \times 10^{-1}$ $pg_{1\text{-}NPYR}\ mg^{-1}\ min^{-1}\ [NO_2]$ could be obtained.

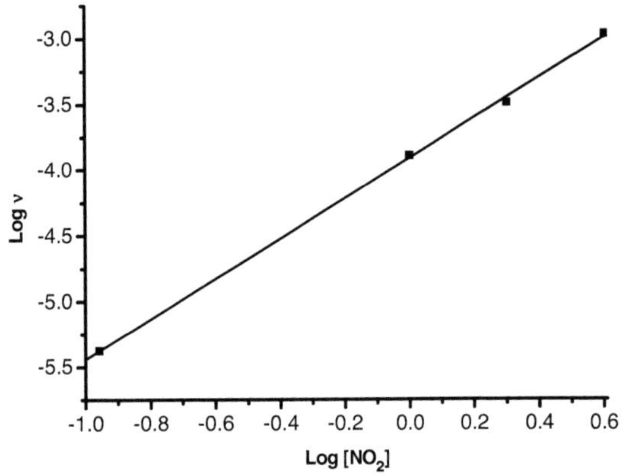

Figure 4-13. Linear least square fit of the logarithm of the data points in Figure 4-11 (equation 3).

4.2.2 Nitration reaction kinetics: BaP-coated soot

In regards to BaP coated soot, the influence of the reaction time on the nitration reaction was studied in the presence of 2.5 ppm and 5 ppm (m/m). BaP was more reactive than PYR towards nitration, as already found in previous studies (114). The main reaction product was 6-NBaP with trace amounts of 1- and 3-NBaP. Formation of 6-NBaP increased with increasing reaction time. In the experiments at 5 ppm, it increased until 180 min while showing an apparent decrease at higher reaction time, as can be seen in Figure 4-14. The data points are characterized by a high standard deviation. The reason for the high variation of the 6-NBaP production was the high variability of the BaP coating due to the high temperature at which it was necessary to coat the particles. Since the BaP was so reactive towards nitration, at reaction time > 1 hour the initial amount of BaP was the limiting factor to determine the amount of 6-NBaP. Such a high dispersion of the data points did not allow a proper fitting, making difficult the interpretation of some features like the apparent decreasing of 6-NBaP amount at a reaction time of 300 min. This decrease could be explained by the further reaction of 6-NBaP with NO_2 to form dinitro—BaP like 1,6- or 3,6-DNBaP. Unfortunately those substances were not found.

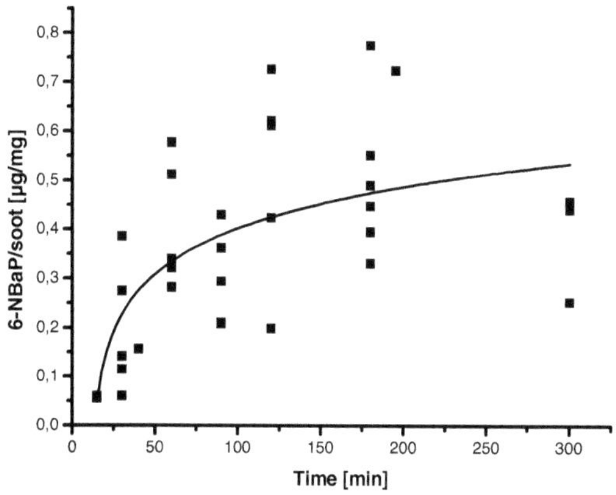

Figure 4-14. Normalized amount of 6-NBaP on soot aerosol versus the reaction time. The curve is a non linear least square fit.

For this reason, the results has been plotted as the conversion of BaP to 6-NBaP versus reaction time, as can be seen in Figure 4-15 both for the 5 ppm and 2.5 ppm series. Data points and error bars represent means and standard deviations of three to six repeated measurements performed at 25°C.

The amounts of 1- and 3-NBaP were negligible in comparison to 6-NBaP, therefore under the assumption that the nitration was the only ongoing reaction, the conversion was calculated as follows (132)

$$\text{Conversion [\%]} = \frac{[\text{6-NBaP}]}{[\text{BaP}] + [\text{6-NBaP}]} \cdot 100,$$

where [BaP] and [6-NBaP] are the normalized amount on soot aerosol after the nitration reaction. The decrease at reaction time = 300 min was then simply due to a lower BaP amount that had been coated on the soot particles. As already mentioned, BaP showed a higher reactivity towards nitration than PYR. With $[NO_2]$ = 5 ppm, at a reaction time of 180 min 80% of BaP was converted to 6-NBaP. The logarithmic increase of the 6-NBaP production is compatible with the linear dependence of the 1-NPYR formation with the time observed in PYR coated soot. In that case PYR conversion after 3 hours (180 min) was only 10%, and can be considered as the initial phase of the reaction where PYR is not the limiting factor.

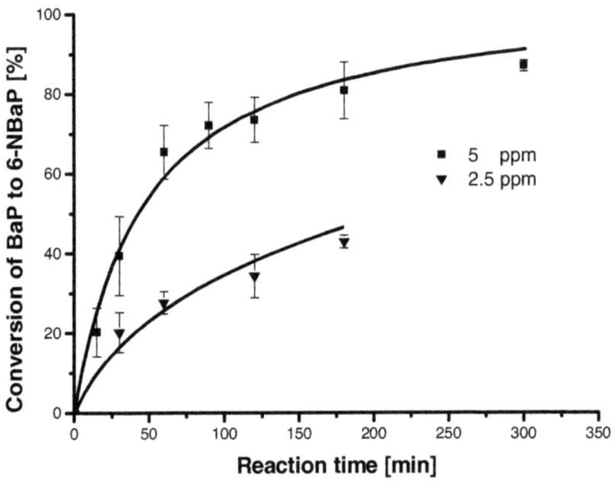

Figure 4-15. Conversion of BaP to 6-NBaP. The data points and error bars represent mean and standard deviation of three to six repeated measurements performed at 25°C. The curves are nonlinear least-square fits.

The results obtained with PYR- and BaP-coated soot exposed to NO_2 under diesel exhaust sampling conditions show that a quantifiable amount of NPAH is produced. This is of crucial importance for the determination of artifacts during diesel soot sampling. Besides, if enough reaction time is provided, nitration could also occur before the particles leave the tailpipe of the vehicle. This could be the case with diesel exhaust aftertreatment systems like DPFs. However, the experiments described above, hardly simulate real DPFs conditions because of the low temperature (20°C) at which they were carried out. The effect of the temperature may play a key role and will be investigated further.

4.2.3 Effect of temperature

The effect of reaction temperature on the reaction rate of the 1-NPYR formation was investigated at a reaction time of 30 min and under a NO_2 concentration of 2 ppm at four different temperatures ranging from 20°C to 300°C. Results are plotted in Figure 4-16. Data points represent the mean and standard deviation of four replicate measurements.

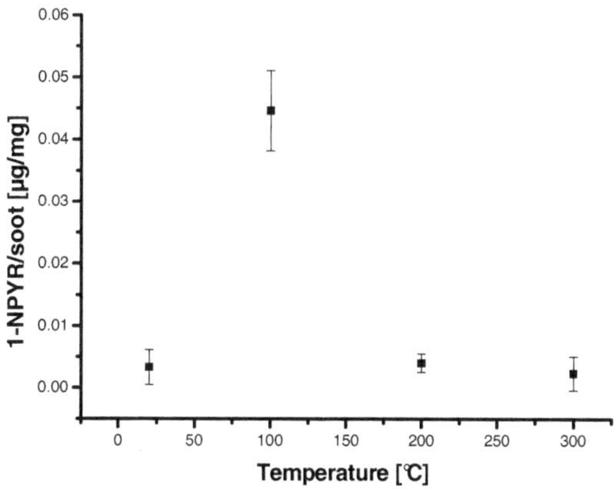

Figure 4-16. 1-NPYR normalized to soot amount formed versus reaction temperature.

The 1-NPYR formation found on soot particles was highest at 100°C. At higher temperature 1-NPYR content was at the same level as found at 20°C.

The decrease of 1-NPYR found on soot particles at T > 100°C does not necessarily indicates an impediment of the nitration reaction. More probable at such temperatures is the thermal desorption of PYR or of the newly formed 1-NPYR; it may be that 1-NPYR desorbs into the gas phase and thus becomes undetectable on the aerosol particles.

This is of importance for real DPFs operation conditions because at the temperature at which these systems work (200°C – 500°C), a significant fraction of the NPAH could be found in the gas phase. Furthermore, our experiments with PYR- and BaP-coated soot showed both a significant

NPAH formation at 20°C. The hypothesis of a cold start of the engine with NPAH formation followed by desorption at higher temperature is plausible.

To examine these points, gas/solid-phase experiments were carried out. Quartz fiber filters were spiked with 10 ng of 1-NPYR and 10 ng of 6-NBaP and then heated up to 150°C and 250°C for 30 minutes under a flow of nitrogen of 1.5 L/min.

As described in the experimental section, the gas phase was sampled by two subsequent porous polyurethane foam collectors (PUF) placed in line. As can be seen in Figure 4-17, after thermal treatment at 150°C, less than 5% of 1-NPYR and 32 ± 13% of 6-NBaP could still be found on the filter, while respectively 54 ± 9.5 % and 52 ± 7.0% were detected on the first PUF and none on the second PUF. At 250°C the fraction of NPAH detected on the filter was lower than 2% both for 1-NPYR and 6-NBaP. In the gas phase, 47 ± 12% of 1-NPYR and 62 ± 11% of 6-NBaP were detected on the PUF. A reason for the incomplete recovery of the analyte could reasonably be a loss during the sampling or incomplete extraction of the PUFs.

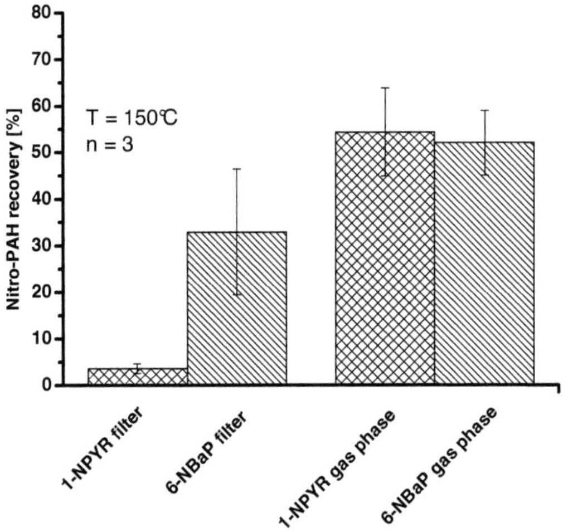

Figure 4-17. Gas/solid-phase partitioning of 1-NPYR and 6-NBaP at 150°C. Data and error bars represent the mean and standard deviation of three repeated measurements.

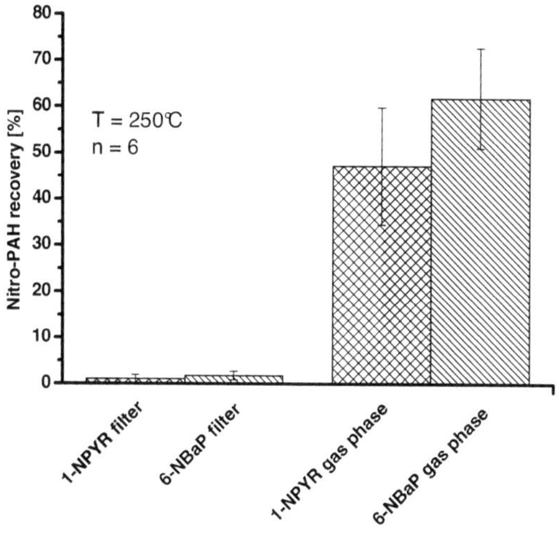

Figure 4-18. Gas/solid-phase partitioning of 1-NPYR and 6-NBaP at 250°C. Data and error bars represent the mean and standard deviation of six repeated measurements.

4.2.4 DPF simulation system: PM-Kat®

The nitration reaction of BaP-coated soot has been investigated also in a DPF simulation system which was designed to trap particles using the deep-bed filtration mechanism as it happens in PM-Kat® traps. This system, described in section 3.3.2, has a filtering efficiency of 40% measured by SMPS and allows artefact-free sampling at three positions: a) DPF; b) quartz fiber filter after DPF; c) washing flask for the gas phase collection. The influence of the reaction temperature on the nitration reaction was studied with $[NO_2]$ = 120 ppm. Reaction time was 1 hour, according with the capacity of the NO_2 denuders, which efficiently remove NO_2 before the sampling points b and c.

Results are shown below. In Figure 4-19 the amount of produced 6-NBaP and 1-NBaP ,normalized to the soot amount deposited within the DPF structures (sampling point a), is plotted versus reaction temperature. Data points represent the mean and standard deviation of three replicate measurements performed at 25°C, 150°C and 250°C.

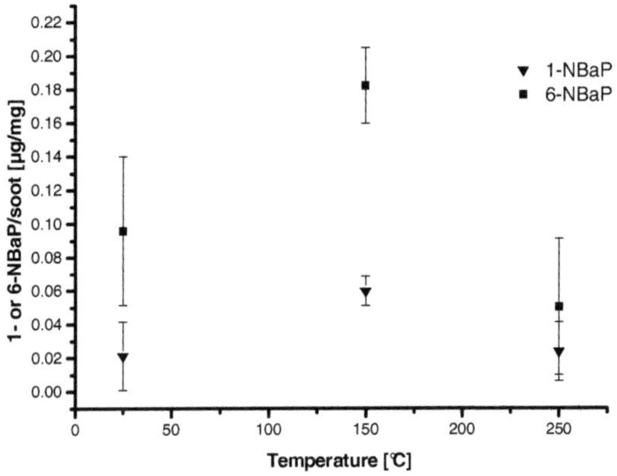

Figure 4-19. Normalized amount of 1- and 6-NBaP on soot collected within the DPF structure. Data and error bars represent the mean and standard deviation of three repeated measurements at 25°C, 150°C and 250°C.

The formation of 6-NBaP within the DPF structures is evident. Besides 6-NBaP, considerable amounts of 1- and 3-NBaP have been found, which represent 15% - 25% of the total NPAH production.

The chromatographic method with fluorescence detection used for the analysis of those samples did not allow the separation of 1- and 3-NBaP, which co-eluted. Therefore it was not possible to differentiate between 1-and 3-NBaP and the results have to be interpreted as sum of the two compounds. However, position 1- is much more prone to nitration than position 3- (196), therefore 1-NBaP should represent the largest amount in comparison to 3-NBaP. For simplicity and illustration purpose, the sum will be considered as if it were only 1-NBaP.

The NPAH formation found on soot is highest at 150°C with 0.18 ± 0.02 μg/mg of 6-NBaP and 0.06 ± 009 of 1-NBaP, while it decreases at 250°C to levels similar to those noticed at room temperature. The trend of 6-NBaP and 1-NBaP is similar, thus supporting the accuracy of the measurements. Single measurements were also taken in the proximity of the maximum. At 175°C amounts of 6-NBaP and 1-NBaP of 0.16 μg/mg and 0.038 μg/mg, respectively, fit really well into the graph in Figure 4-19, according with a maximum at 150°C. At 125°C 6-NBaP and 1-BaP were respectively 0.097 μg/mg and 0.016 μg/mg, in good accordance too.

To assure that the samples, which showed the highest NPAH concentrations did not were casually also those with the highest BaP coated amount, the conversion of BaP into NPAH was calculated too. Results are shown in Figure 4-20, where the amount of BaP that was converted into NPAH (6-NBaP + 1-NBaP) is plotted versus reaction temperature. Data points represent the means and standard deviations of three replicate measurements performed at 25°C, 150°C and 250°C. Indeed at 150°C the BaP conversion to its nitro derivates 1- and 6-BaP is the highest, confirming the results shown before.

In Figure 4-21, the analysis of soot collected on the quartz filter placed after the DPF (sampling position b) is reported. This soot is the fraction which has not been captured from the DPF and/or which has eventually been re-emitted from the DPF. In the first case the reaction time is relatively short and it is the time that the particles need to travel from were NO_2 is added (before DPF) to the where NO_2 is removed (three-channel annular denuder). This time frame is approx. 10 seconds.

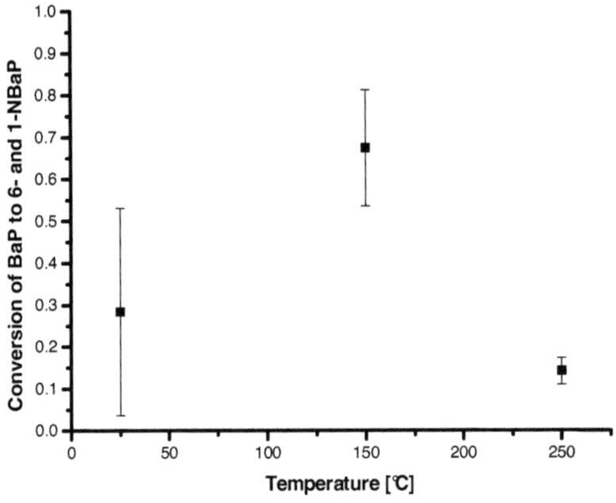

Figure 4-20. Conversion of BaP to 6-NBaP and 1-NBaP. The data points and error bars represent mean and standard deviation of three to repeated measurements performed at 25°C, 150°C and 250°C.

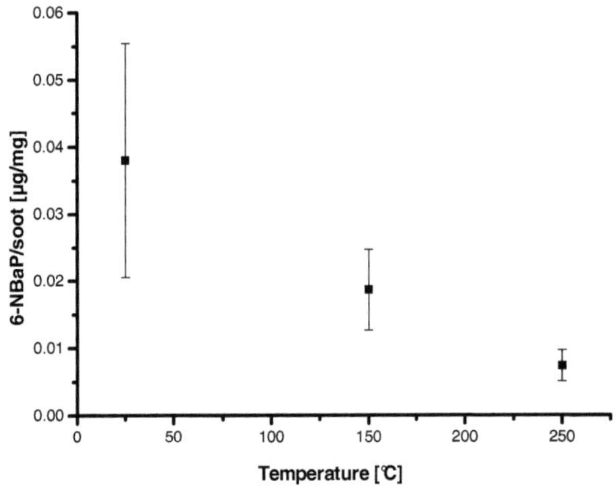

Figure 4-21. Normalized amount of 6-NBaP on soot collected on the quartz filter after the DPF (sampling point b). Data and error bars represent mean and standard deviation of three repeated measurements.

In the second case the reaction time is unknown because it is not possible to know when the particles from the DPF; it then ranges from few seconds (particles re-emitted shortly after interception) to 1 hour (particle intercepted at the beginning of the experiment which became re-emitted at the end).

At this point (sampling position b) only 6-NBaP could be detected, while 1-NBaP was below the limit of detection. Therefore the contribution of the re-emitted particles should be low, otherwise also 1-NBaP should have been found. Moreover, position 6- is much more prone to nitration than position 1-, thus the production of 6-NBaP is favored. 6-NBaP concentration decreases with increasing temperature. A similar trend has been noticed also for BaP (Figure 4-22) found on soot particles. Actually, at 25°C the amount of BaP found on particles is 0.45 ± 0.23 µg/mg, in good accordance with the amount that had been coated, and decreases at higher temperature. Since PAH oxidation tests excluded BaP degradation under such conditions, a possible explanation for this decrease, could be the thermal desorption of both: product and educt.

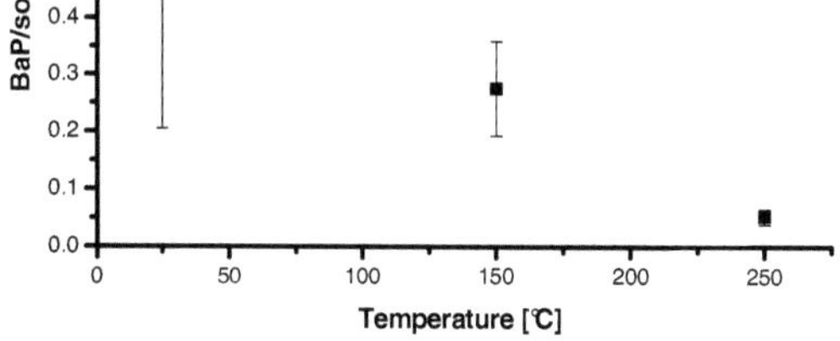

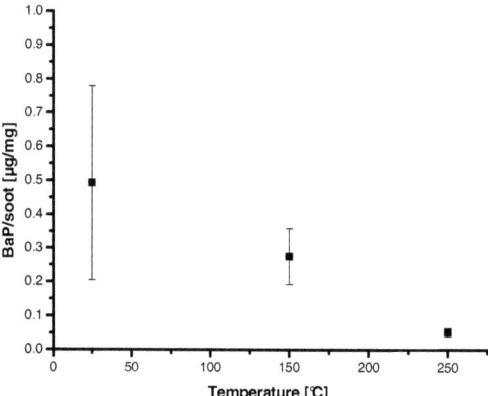

Figure 4-22. Normalized amount of BaP on soot collected on the quartz filter after the DPF (sampling point b). Data and error bars represent the mean and standard deviation of three repeated measurements.

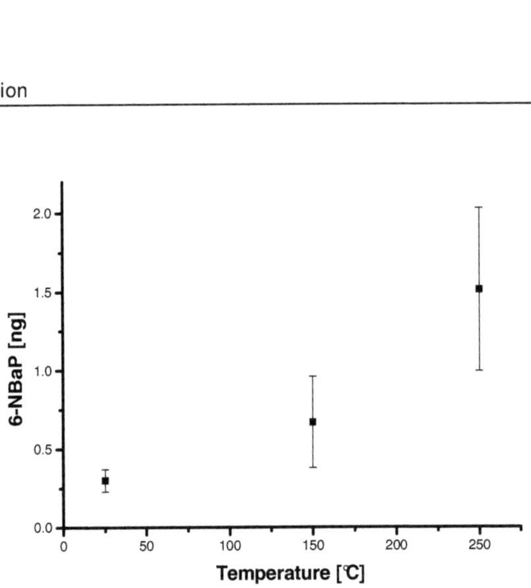

Figure 4-23. Amount 6-NBaP on soot collected in the gas. Data and error bars represent the mean and standard deviation of three repeated measurements.

Data in Figure 4-23 support this hypothesis, since an increasing amount of 6-NBaP with increasing temperature is found in the gas-phase collected by the washing flask. This could originate from the thermal desorption of the 6-NBaP after its formation, or from reaction that occurs directly in the gas-phase between BaP and NO_2.

Summarizing, since the inlet aerosol was NPAH-free, on the basis of these results, one can affirm that the DPF structure supported the nitration chemistry of BaP with the formation of 6-NBaP and 1-NBaP. To find the reaction products by extraction of the soot deposited within the DPF structure (sampling point A), demonstrates that a part of the reaction occurs heterogeneously on the soot surface between adsorbed BaP and NO_2. The amount of NPAH found is highest at the sampling point A in comparison to sampling points B and C. The reason is that those particles had more time to react since they were trapped within the DPF and become nitrated form the NO_2 passing through it.

Heterogeneous reaction could occur also on particles which were not intercepted from the PM-Kat® structures. In this case particles traveled directly to sampling point B and the reaction time with NO_2 was short. Since position 6- is much more prone to nitration than positions 1- and 3-, its production was favored and only 6-NBaP was detected.

By increasing the temperature in the DPF an increasing amount of 6-NBaP was found in the gas phase, showing a thermal desorption of the reaction product. This is of a certain importance for real DPFs system where the high temperature could cause a sublimation of the NPAHs which would be emitted in the gas phase and only after cooling and dilution in the air would eventually be adsorbed again on soot particles. That is even more relevant for PAH (211), which have an even lower vapor pressure than their nitroderivates, and can be found either in the gas-phase or in the particulate phase depending on their structure an engine load (212). Therefore BaP could sublimate into the gas phase making it unavailable for the nitration on the soot particles. For this reason, the production of 6-NBaP and NPAH in general could occur directly in the gas phase, but in the present dissertation the focus is in a first step on the generation and release of NPAH from the deposited (and the already PAH-loaded) aerosol within a trap. At our institute, following the indications emerged from the present work, other experiments have begun to investigate the NPAH formation in the gaseous state elucidating this relevant aspect.

4.3 Effect of DPF on heavy duty vehicle PAH and NPAH emissions

Besides the laboratory investigation on NPAH formation under DPF relevant conditions, the sampling and the analysis of real diesel soot was considered of central importance for the completeness of the study. It is worth to remind that this study was independently founded by the German Research Foundation. Because of that, no car manufactures were present in the project as official partner.

For the sampling and analysis of diesel soot from real engines and vehicles, the Institute of Combustion Engines (TUM), BMW Group AG and MAN AG has been contacted. The latter was the only that accepted the cooperation. For this reason, one heavy duty engine could be investigated with focus on the effect of DPF, while it was not possible to get samples from passenger cars.

Sampling conditions and analytic procedure are explained in detail in section 3.4.2. At this point is noteworthy to remind that two different operating points of the ESC cycle have been sampled: B25 (low load, medium engine speed) and C100 (high load, high engine speed). The three sampling positions were:

A Raw (before oxi-cat)

B Before DPF (after oxi-cat, before DPF)

C After DPF

The DPF used was a wall-flow filter passively regenerated by the NO_2 produced by an oxidation catalyst placed in front of it. NPAH artifact-free samples were assured by scrubbing the NO_2 with 2 annular denuders placed in front of the sampling filter. After sampling and weighing, filters were extracted and HPLC analysis was carried out. The next two sections report and discuss in detail the results obtained in regard to PAH and NPAH emissions.

4.3.1 PAH emission

Results regarding PAH emissions normalized to the exhaust gas mass, which is expressed in Kg, are shown in Figure 4-24. The columns (with the exception of column "regeneration", which will be discussed later) represent the average of three to five measurements taken in two different sampling campaigns at MAN test benches in Nürnberg (Germany).

The highest total PAH emission was 4416 ng of PAH / kg of exhaust gas at sampling position A and operating point B25 (B25-A). In this case the most emitted PAHs were phenanthrene (35% of the total emissions), pyrene (24%), naphthalene (11%) fluoranthene (10%) and benzo(a)anthracene (9%). Among the 15 PAHs that have been analyzed, those 5 represented more than 85% of the total emissions at all sampling points. Because of that, for illustration purposes in Figure 4-24 the other 10 PAHs are indicated as "other PAHs".

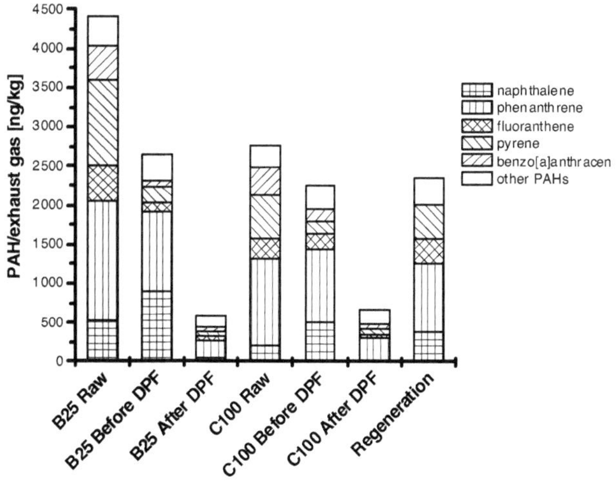

Figure 4-24. PAH emission normalized to the exhaust mass.

Table 4-2. Total PAH emission (normalized to exhaust gas mass expressed in kg). Emission at sampling point A is considered 100%.

Sampling point	Total PAH emission [ng/mg]	Total PAH emission [%]
B25-A (Raw)	4416	100
B25-B (Before DPF)	2636	65
B25-C (After DPF)	575	13
C100-A (Raw)	2755	100
C100-B (Before DPF)	2240	81
C100-C (After DPF)	653	23

PAH emissions of operating point B25 were generally higher than those of C100, with the exception of sampling point C, were C100 shows a slightly higher emission in comparison to B25 (653 ng/kg in comparison to 575 ng/kg). Nevertheless, the reduction of total PAH emission along the aftertreatment system is remarkable and is resumed in

Table 4-2.

Emission factors (EF) are listed in Table 4-3. EF are calculated dividing the emissions of each PAH at all sampling points by the emission at sampling point A. Thus, sampling point A has an EF of 1.00. EF lower than 1.00 means a PAH emission reduction (e.g. EF = 0.10 means an emission of 10% of that at sampling point A) while EF higher that 1.00 indicates de-novo PAH formation (e.g EF= 2.00 signifies a doubling of the emission). EF of 0.00 means no emission. The PAH emission profile changed along the aftertreatment system showing that not all PAH emissions became reduced with the same efficiency. The difference could also be addressed to a different gas phase/particulate partition between different PAH, because of the decreasing temperature along the tailpipe. Because only the particulate phase was sampled, there is no possibility to know to which extend PAHs were also present in the gas phase. However, before sampling exhaust gas became diluted and cooled down to the sampling filter temperature of 25°C. Therefore we presume that the greatest part of gas-born PAHs were already condensed again on the particulate phase.

Table 4-3. PAH emission normalized to the exhaust mass at sampling position A (raw exhaust). Emission factors (EF) at sampling points B and C. EF account for the increase or decrease of PAH emitted in relation to sampling position A both at operating points B25 and C100.

PAH	B25 –A [ng/kg]	B25 –B [EF]	B25 –C [EF]	C100 –A [ng/kg]	C100 -B [EF]	C100 –C [EF]
Naphthalene	514.5	1.74	0.11	203.6	2.48	0.06
Acenaphthene	5.5	5.42	9.68	5.3	1.28	16.5
Fluorene	151.9	1.22	0.26	99.2	1.91	0.43
Phenanthrene	1534.7	0.66	0.14	1120.1	0.83	0.25
Anthracene	139.0	0.62	0.05	136.9	0.54	0.11
Fluoranthene	464.1	0.27	0.12	248.1	0.78	0.23
Pyrene	1079.2	0.19	0.05	548.3	0.31	0.13
Benzo[a]anthracene	435.1	0.18	0.13	371.1	0.41	0.14
Chrysene	20.2	0.47	0.35	5.7	1.85	0.93
Benzo[b]fluoranthene	52.5	0.04	0.00	5.4	0.79	0.09

Benzo[k]fluoranthene	1.3	1.51	1.38	1.0	1.63	0.89
Benzo[a]fyrene	14.2	0.58	1.60	10.2	0.38	1.95
Dibenz[ah]anthracene	0.0	0.01	0.00	0.00	0.00	0.00
Benzo[ghi]perylene	4.0	0.23	2.10	0.00	0.00	0.00
Indeno[1,2,3,c,d]pyrene	0.0	0.00	0.00	0.00	0.00	0.00

Among the 5 major PAHs emitted, naphthalene and pyrene were those with the highest reduction (lowest EF at sampling points B and C). Results like the reduction efficiency of pyrene > benzo(a)anthracene > fluoranthene > chrysene, nicely agree with existing iterature data (35). While EF of almost all investigated PAHs decreased, emission of acenaphtalene and benzo(a)pyrene at sampling points C were found to be higher than at sampling point A (EF>1). This could indicate a de-novo formation due to soot decomposition along the aftertreatment system and within the DPF in particular. EF for benzo(a)pyrene are 1.60 and 1.95 for C-B25 and C-C100 respectively. Because of that, the contribution of benzo(a)pyrene to the total PAH emission increased form 0.3% at sampling point A to 3.2% at sampling point C (after DPF).

The column "regeneration" is the result of the analysis of one single measurement sampled while the filter was actively regenerated by mean of fuel post-injection. In fact, the DPF used regenerated passively by the NO_2 passing through it, following the CRT principle. However, that system had also the possibility to switch to active regeneration with fuel post-injection. To investigate the PAH and NPAH emission during this kind of regeneration mode, the DPF was loaded with approx 80 mg of soot by operating the engine at low load and low engine speed. Under these conditions the temperature in the DPF was below 300°C and the passive regeneration rate was at its minimum. Because of that, the rate at which soot was deposited in the DPF was higher than that at which it was oxidized, thus getting a net increase of the soot mass in the DPF. After that, the filter was actively regenerated by fuel post-injection. The temperature raised to 640°C and the filter was fully regenerated within 15 minutes. At this temperature the $NO \leftrightarrow NO_2$ equilibrium is almost completely shifted to NO, therefore soot was oxidized only by oxygen, which was 9% in volume. Because of time restrictions, it was possible to sample the emission of this regeneration mode only once and only at sampling point C. The engine operating point was B25, thus this measurement can be also considered as a B25-C with active regeneration.

Total PAH emissions of the active regeneration mode were 2347 ng/Kg, about four times higher than those of B25 with passive regeneration at the same sampling point. As can be seen in Figure 4-24, the PAH emission profile is slightly different from B25. Phenanthrene represents 37% of the total emissions and is again the most emitted PAH followed by pyrene with 18% and naphthalene with 17%. Acenaphthene is not emitted. In addition, compounds like benzo(a)pyrene and benzo[ghi]perylene which were scarcely present in other samples, were emitted with 75 ng/Kg and 55 ng/Kg, respectively 3.2% and 3.5% of the total emission. These results indicate that a part of the found PAHs may not come from the engine. They can rather originate from the combustion of the fuel which was used for the regeneration. Fuel post-injection occurred just before the DPF. Fuel droplets burn in a completely different environment than that of the cylinder. Therefore it is reasonable to make the hypothesis that the PAH originated from combustion under such conditions are different from those produced from combustion in the engine, thus changing the PAH emission profile.

PAH emissions have been also normalized to the soot mass, which was collected on the sampling filter. These results are shown in Figure 4-25. In contrast to PAH emission normalized to the exhaust mass, in this case emissions of operating point C100 are higher than those of B25 at all sampling positions. This agrees with literature data, where a higher amount of particulate bond PAHs at high load in comparison to low load and idle was found (212).

Of course the PAH emission profiles do not change in comparison to the exhaust-normalized data. The percentage contribution of singles PAH to the total emission remains the same. Nevertheless, the reduction efficiency, even if still evident, is less pronounced. Data are resumed in Table 4-4. The PAH degradation occurs with high rate already in the oxidation catalyst and before DPF. In addition, the lower reduction efficiency encountered when data are normalized to the soot amount, indicates that soot becomes oxidized faster than PAHs. For instance, PAH emissions of operating point B25 at sampling point B and C are respectively 58% and 54% of raw exhaust emissions.

Table 4-4. Total PAH emission (normalized to the soot amount expressed in mg). Emission at sampling point A is considered 100%.

Sampling point	Tot PAH emission [ng/mg]	[%]
B25-A (Raw)	108.4	100
B25-B (Before DPF)	63.70	58
B25-C (After DPF)	59.30	54
C100-A (Raw)	188.8	100
C100-B (Before DPF)	98.22	52
C100-C (After DPF)	78.08	41

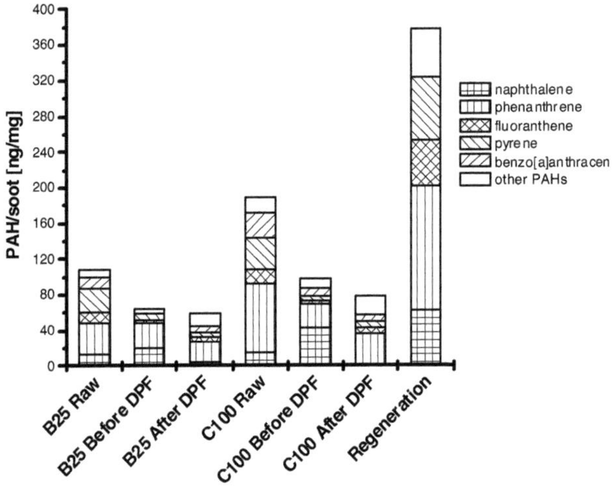

Figure 4-25. PAH emissions normalized to the soot amount.

This means that to a consistent decrease of the soot mass after the DPF, does not correspond a parallel decrease of the PAH amount. Furthermore, it is also plausible that the Pt coating in the oxidation catalyst plays a role in PAH degradation, which occurs more slowly within the uncoated structure of the cordierite-based DPF. Regarding to PAH emissions during the active regeneration mode, in this case was produced the soot with the highest PAH content (377 µg/mg). The high ratio is due to the small denominator, confirming again rapid soot combustion and slower PAH oxidation. The last column of Figure 4-25 that should be discussed is the column called "DPF soot". It represents the PAH concentration of soot which was directly obtained by putting the DPF in vertical position, beating it with a stick and collecting that soot that came out. This operation was made at the end of the sampling campaign after the last operation point of the program, that is a C100. As can be seen, that soot is relative poor of PAH and with 20 ng/mg it is the soot with the lowest PAH content. A possible explanation could be that at the temperatures found in DPF during the last C100 cycle (430°C - 440°C) most of the adsorbed PAH sublimated into the gas-phase leaving a relatively "clean" soot trapped in the DPF. Naphthalene was the most abundant PAH found. Having the lowest vapour pressure among all investigated PAHs it is surprising to find it in such an amount. It could be that naphthalene, as well as others PAH, have been incorporated during the soot formation process, and not being on the surface could not sublimate because "trapped" within the soot structure. Only after sonication with proper solvents it could be extracted and analysed.

4.3.2 NPAH emissions

The same soot samples extracted for the PAH analysis, were used also for the quantification of the NPAH emissions. The operating points sampled were again B25 and C100. In Table 4-5 are reported the exhaust gas temperature and the NO_2 concentration measured during the tests.

Table 4-5. Exhaust gas temperatures an [NO_2] at different sampling positions and operating points.

Sampling point	NO_2 [ppm]	Temperature [°C]
B25-A (Raw)	5-8	335
B25-B (Before DPF)	130-140	315
B25-C (After DPF)	120-130	300
C100-A (Raw)	8-12	470
C100-B (Before DPF)	90-100	450
C100-C (After DPF)	30-75	440

Engine-out NO_2 concentrations were considerably lower than those measured downstream the aftertreatment system. That is due to the action of the oxidation catalyst, which converts the engine-out NO to NO_2 which is needed for the regeneration of the DPF. [NO_2] at sampling position B and C were higher by operating the engine at point B25 than at C100. That can be explained by to two effects. The first is the lower catalytic conversion efficiency at point C100 than at point B25, mainly because of the higher temperature and the lower oxygen content at high load than at low load. The second effect is the already mentioned NO $\leftrightarrow$ NO_2 equilibrium, which at high temperature shifts to NO. The decrease of the NO_2 concentration after the DPF ($[NO_2]^C$) in comparison to the concentration measured between Oxi-cat and DPF ($[NO_2]^B$), is due to the NO_2 consumption caused from soot combustion within the DPF. This is a good indicator to follow the state of the passive filter regeneration. With the aftertreatment system tested, applied to a HD engine, it was empirically determined that a difference between $[NO_2]^B$ and

$[NO_2]^C$ of < 15 ppm, indicates a slow regeneration or a small amount of trapped soot. On the other hand, a difference > 30 ppm indicates a fast regeneration and a highly loaded DPF. The latter case happened always when the operating point C100 was driven after one or more B25, indicating lower passive regeneration rates at low load than at high load.

Concerning NPAH emission, beside those formed during combustion in the engine, others can form from precursor PAHs via nitration in the gas or particulate phases along the aftertreatment system. Many nitrating species such as $\bullet NO_2$ or $\bullet NO_3$ radicals are present in the exhaust gas. Moreover, also HNO_3 and HNO_2 can be present originating the strong electrophilic species NO^+ and NO_2^+. Hydroxy radicals (•OH) can attack PAHs in a first step, followed by the addition of nitrating species (35,213). The high temperature could also favor the direct nitration of NO_2 radicals, which would compete with the •OH attack (208). Therefore a DPF, providing enough reaction time, could support all these processes. That would be the case of heterogeneous nitration reaction, which could occur on particles trapped within the filter structure. Our experiments in the DPF simulation system show that even at 250°C one does not have to expect "naked" soot particles since there are enough strong sorption-active sites on a particle where a PAH molecule may become adsorbed. Therefore, the reaction can occur on the soot surface. On the other hand, many PAHs can be already gas born and others can volatilize from the particulate phase into the gas phase. The eventual nitration reaction could then occur directly in the gas-phase. In this case the filter structure itself would not have any influence on the reaction time, which would be relatively short. Despite of that, a CRT like the one investigated in the present study, produces high $[NO_2]$ in the oxidation catalyst to regenerate the DPF. Thus, the gas-phase nitration reactions could proceed much faster because of the enhanced NO_2 concentration. Additionally, also a further oxidation of engine-out PAH and NPAH may be caused by higher NO_2 concentration (35). The artifact-free NPAH sampling and the NPAH analysis carried out in this study allowed to show which of these competing processes prevailed.

Results of NPAH emissions normalized to the exhaust gas mass expressed in kg, are reported in Figure 4-26. NPAH emission levels are 10 - 40 times lower than those of PAH and range from 10 ng/kg to 131 ng/kg. Concerning the effect of the aftertreatment system, reduction of the total

NPAH emission was observed. Nevertheless substantial differences emerged between the two operating point B25 and C100 and between different sampling positions.

As for PAH before, NAPH emission reductions are referred to sampling point A (raw exhaust) which showed the highest NPAH emissions at both operating point B25 and C100 with 131 ng/kg and 91 ng/kg respectively. Regarding the operating point C100, after the DPF were emitted only 10% of the NAPHs found at sampling point A (Table 4-6).

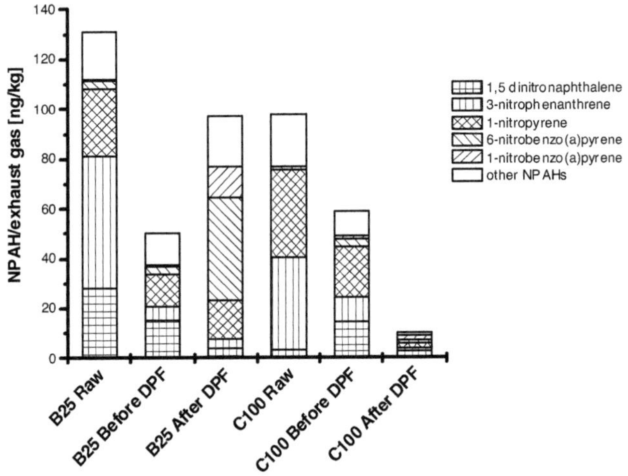

Figure 4-26. NPAH emission normalized to the exhaust mass.

The NPAH emission profile of raw exhaust is quite similar at both operating point B25 and C100. 3-nitrophenanthrene (53.03 ng/kg and 37.14 ng/kg) is the most emitted compound followed from 1-nitropyrene (27.14 ng/Kg and 35.26ng/kg). The main difference between the two points, is the emission of nitro-derivatives of naphthalene, which are sensibly lower at C100 than at B25. The origin of this difference is uncertain. One explanation could be the higher temperature of C100 in comparison to B25 having an effect on the relatively volatile compounds like nitro-naphthalenes. However, the reduction of the total NPAH emission along the aftertreatment system by operating the engine at point C100, and the doubling of the emission between sampling position B and C at operating point B25, are particularly evident.

Table 4-6. Total NPAH emissions (normalized to exhaust gas mass). Emission at sampling point A is considered 100%.

Sampling point	Tot NPAH emission [ng/Kg]	[%]
B25-A (Raw)	131	100
B25-B (Before DPF)	55.5	38.4
B25-C (After DPF)	97.0	73.8
C100-A (Raw)	97.7	100
C100-B (Before DPF)	58.8	65.1
C100-C (After DPF)	9.93	10.1

EF for NPAH have been calculated in the same way as for PAH, that is by dividing the emission at all sampling points by the emission at sampling point A. EF are listed in Table 4-7. 3-nitrophenanthrene, which at sampling point A was the most emitted NPAH in both B25 and C100 operating points (40% and 38% of the total emissions respectively) showed high reduction with EF of 0.08 and 0.03 at sampling point C. Also the mutagenic 1-nitropyrene, which was the second most emitted NPAH, was significantly reduced along the aftertreatment system, especially at operating point C100. After the DPF, its emission was only 5% of the emission found in raw exhaust. Among the 11 NPAH analyzed, 8 have an emission factor <1 or even show no emission (EF = 0). Despite of that, emissions of three compounds increased. They are 1,6-dinitropyrene, 6-nitrobenzo(a)pyrene and 1-nitrobenzo(a)pyrene. 1,6-nitropyrene was found neither at sampling point B, nor in raw exhaust, indicating a new formation within the DPF structure. This is of particular importance since 1,6-dinitropyrene is considered a supermutegen. The other two, both strong mutagenic, increased their emissions by a factor 1.59 and 5.95 respectively.

Table 4-7. NPAH emission normalized to the exhaust mass at sampling position A (raw exhaust). Emission factors (EF) at sampling points B and C. EFs account for the increase or decrease of NPAH emitted in relation to sampling position A both at operating points B25 and C100.

NPAH	B25 –A [ng/kg]	B25 –B [EF]	B25 –C [EF]	C100 –A [ng/kg]	C100 -B [EF]	C100 –C [EF]
1,5-dinitronaphthalene	27.8	0.53	0.14	3.3	4.32	0.83
1-nitronaphthalene	2.5	0.70	0.52	2.9	1.29	0.00
2-nitronaphthalene	0.1	4.11	19.7	1.3	0.97	0.61
2-nitrofluorene	0.8	0.07	0.98	2.5	0.23	0.00
3-nitrophenanthrene	53.0	0.11	0.08	37.2	0.27	0.03
1,6-dinitropyrene	0.0	-*	-*	0.0	0.00	-*
1-nitropyrene	27.1	0.48	0.56	35.3	**0.58**	0.05
3-nitrofluoranthene	2.5	0.00	2.14	3.8	0.23	0.04
7-nitrobenzo(a)anthracene	13.3	0.81	0.53	10.4	0.33	0.00
6-nitrobenzo(a)pyrene	3.4	0.81	12.2	0.8	4.02	1.59
1-nitrobenzo(a)pyrene	0.7	**1.03**	16.5	0.3	3.35	5.95

* Since no 1,6-dinitropyrene was found at sampling position A, EF could not be calculated. Its emission were: B25-B = 0.13 ng/kg; B25-C = 3.12 ng/kg; C100-C = 0.11 ng/kg

These results are consistent with those observed at operating point B25. Again a reduction in the total NPAH emissions could be observed at sampling position C in comparison to sampling position A. However, emission after DPF are almost double than at position B. That is mainly due to a tremendous increase of 1- and 6-nitrobenzo(a)pyrene emission, which increased by a factor of 16.5 and 12.2, respectively. At operating point B25, 6 NPAH became decreased in their emission and 5 increased. Among others, 2-nitronaphthalene shows a considerable increase (EF = 19.7), but its contribution to total emission remains rather low with 2.9%. 1,6-dinitropyrene, which in raw exhaust was under the limit of detection, was newly formed and it could be detected only downstream the aftertreatment systems. Among compounds whose emissions were reduced in comparison to raw exhaust, many were degraded less efficiently that at C100, showing higher EFs. For instance, 7-nitrobenzo(a)anthracene has an EF of 0.53 and 0.00 at B25 and C100, respectively. Besides, at operating point B25 and sampling point C, all 11 NPAH could be detected and no EF of 0.00 has been reported.

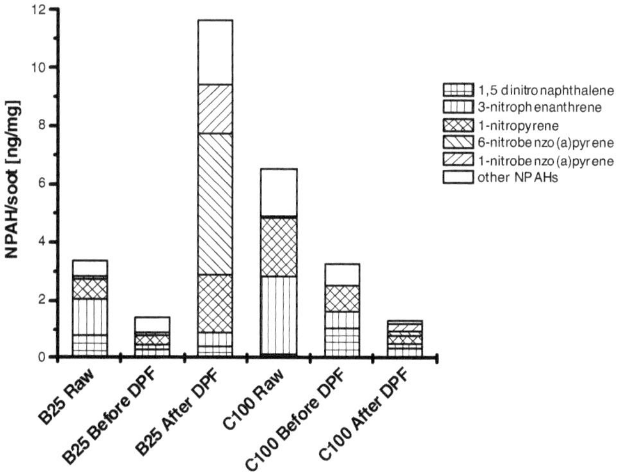

Figure 4-27. NPAH emission normalized to the soot amount.

NPAH emissions have been also normalized to the soot mass expressed in mg, which was collected on the sampling filter. These results are shown in Figure 4-27. Consistent with PAH results, emissions at operating point C100 are higher of those of B25 at sampling positions A and B. Nevertheless, the highest NPAH emission was observed in samples of B25-C. In this case each mg of soot emitted is covered with 11.5 ng of NPAH, a ratio more than 4 times higher than at B25-A. That is of considerable environmental consequence, since an HD vehicles equipped with a CRT, at least at low load and low-medium engine speed, would emit considerably less soot (due to the high soot filtration efficiency), but that soot would be coated with an higher amount of NPAH.

The analysis of soot collected within the filter (see section 4.3.1) gave results similar to those of PAH, that is relatively clean soot. Explanation could be again volatilization of NPAH because of the high temperatures (440°C) in the DPF during the cycle C100 that indeed seemed to degrade and/or volatilize most of the NPAH.

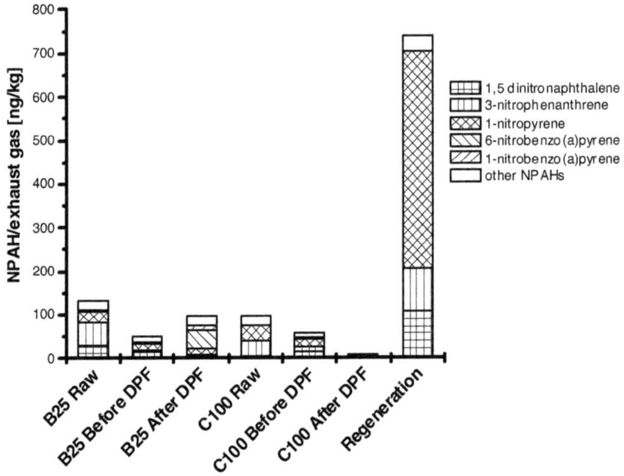

Figure 4-28. NPAH emission (including regeneration) normalized to the exhaust mass.

In regard to NPAH emission during the active regeneration mode (see section 4.3.1), results normalized to exhaust mass are plotted in

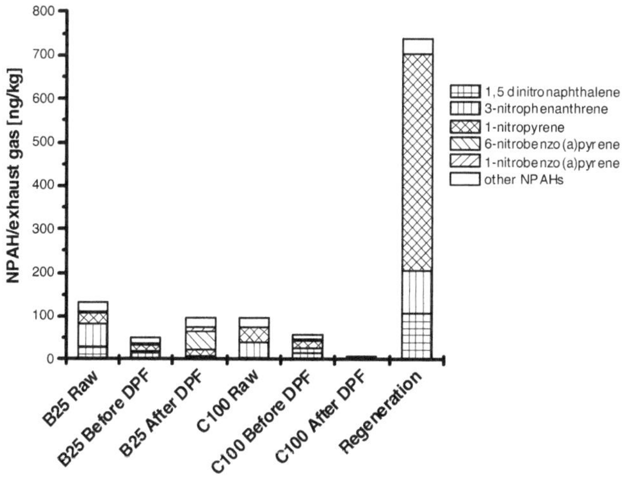

Figure 4-28. As already mentioned, for technical reason it was possible to sample soot from the active regeneration mode only once. As can be seen, the total NPAH emission is much higher than that of all other measurements, which are plotted for comparison. During the active regeneration mode with fuel post-injection, NPAH emissions reached 738 ng/kg and were 7.5 times higher than those of B25 with passive regeneration. Normalizing the emission to soot amount, NPAHs increased by a factor 12.5 (144 ng/mg). The most emitted compound is by far 1-nitropyrene (67% of the total emissions), followed by 3-nitrophenanthrene and 1,5 dinitronaphthalene.

Many of the considerations made for the PAH emission are valid also in this case. In fact, the emission profile is rather different than that of all others samples, confirming the hypothesis of a different origin of the NPAHs in the active regeneration mode. That could be the combustion of the fuel droplets injected just before the Oxi-cat. Artifact NPAH formation on the sampling filter is to exclude both for the use of a efficient NO_2 scrubbing system and for the reason that at 640°C, the temperature at which the active regeneration occurs, few NO_2 is present (< 5 ppm). This fact also supports the hypothesis of a combustion origin rather than nitration of PAH precursors because of the much lower NO_2 concentration during active regeneration than during the passive one.

Summarizing the results regarding NPAH emission from a heavy-duty vehicle equipped with a CRT, it was found that the aftertreatment system tested reduced the total amount of NPAH emitted both at low load and high load. Nevertheless, at low load the lower temperatures and the higher NO_2 concentration could furnish better conditions for nitration, thus shifting the equilibrium NPAH degradation $\leftrightarrow$ NPAH new formation to the right. That was more pronounced for some NPAH as 1- and 6-nitrobenzo(a)pyrene in comparison to others as 3-nitrophanthrene which still became reduced efficiently. For this reasons, the investigation of NPAH emission from CRT applied to passenger cars would be of great interest and should be planned as next step for future investigations. In fact, exhaust gas temperatures of passenger cars are sensibly lower than those of HD vehicles, and that could have an effect on the competitive processes of NPAH degradation vs. NPAH new formation.

The NPAH emission profile changed along the aftertreatment system. For this reason, the average NPAH pattern of diesel exhaust emissions in the atmosphere could change considerably with use of CRTs. In regard to emissions from the active regeneration mode, tremendous high emissions of 1-nitropyrene were reported. More accurate investigation of such a regeneration mode would be of great interest for the future.

4.4 Effect of biofuels on NPAH emissions

NPAH emissions from different vehicles depending on the usage of biofuels were measured within the biofuels project. Samples were taken after dilution tunnels at vehicle roller test benches in Graz and Vienna during defined test cycles. A heavy-duty vehicle (MAN, EURO V), a tractor motor (John Deere, EU III) as well as passenger cars (Audi, diesel, EURO V and Saab, gasoline, EURO IV) were used during this study. To determine differences in the emission rates, fossil diesel, biodiesel and vegetable oil as well as mixtures with different amounts of biofuels in fossil fuel were investigated. Particle bound PAH and NPAHs were sampled on quartz fiber filters, which were extracted and cleaned-up for HPLC-FD quantification. This section summarizes the results of these sampling campaigns concerning the NPAH emissions

4.4.1 EURO V heavy-duty vehicle.

At the Technical University of Graz tests with one HD vehicle have been carried out for the estimation of the effects on NPAH emissions by the use of biofuels. The vehicle was operated with two different driving cycles: FIGE and ESC. The first is a transient cycle, while the latter is stationary and are both described in section 3.4.1. The fuels tested were fossil diesel, fuel blends B7, B10 and B7+3 as well as B100 and vegetable oil (PÖL). The latter was tested as refined (series product), cold-pressed and with the Yelltech 2-tanks system, which allowed to operate the engine at the start and the end phase with fossil diesel fuel and for the remaining cycles time with vegetable oil.

In Figure 4-29 and Figure 4-30, NPAH emissions are shown. As already noticed by the PAH analysis (15), NPAH emissions produced during the transient FIGE cycle are higher than those noticed during the stationary ESC cycle.

During the FIGE cycle, 8 different NPAH could be detected. Total NAPH emissions were comprised between 8 ng/m^3 and 15 ng/m^3. Different fuels produced different emissions but no uniform trend could be noticed.

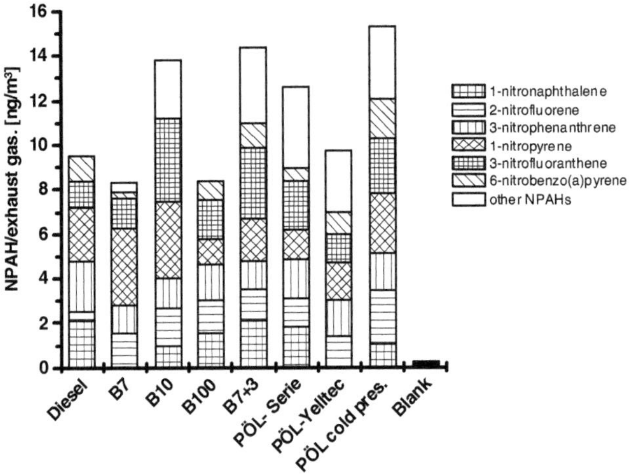

Figure 4-29. NPAH emission from an EURO V HD vehicle (FIGE cycle).

The most emitted NPAH were 2-nitrofluorene, 3-nitrophenanthrene, 1-nitropyrene and 3-nitrofluoranthene. A blank measurement of the dilution tunnel was also executed, and the found NPAH values can be neglected.

Also by operating the vehicle with the stationary ESC cycle (Figure 4-30), no uniform trend could be observed and no correlation with the use of biodiesel or vegetable oil could be found. Emissions ranged between 7 ng/m^3 and 11 ng/mg^3 and again the most emitted compound were 3-nitrophenanthrene, 1-nitropyrene and 3-nitrofluoranthene. Moreover, lower emission of nitro-naphthalenes in comparison to FIGE cycle was noticed.The overall low emission of NPAH from the EURO V HD vehicle can be addressed to the NOx/ soot trade-off effect. In fact, the aftertreatment system applied to the truck was composed only by a SCR system for the reduction of the NOx. Therefore, the engine combustion was optimized for the lowest possible soot emission (high temperatures and no EGR). These conditions could favor the lowering of

PAH and NPAH emissions. The consequent higher NO_x emission could get reduced under the legislation limit by mean of the SCR system.

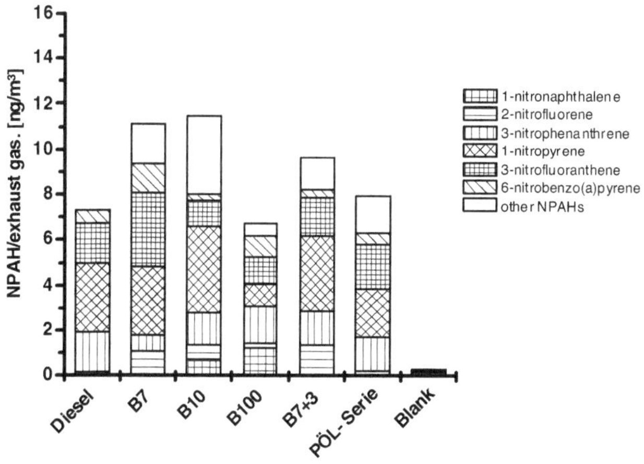

Figure 4-30. NPAH emission from an EURO V HD vehicle (ESC cycle).

4.4.2 Euro V passenger car (diesel engine)

Soot samples for the analysis of NPAH emission from a Euro V diesel-passenger car were sampled at the Technical University of Vienna. Fuels used were fossil diesel, blends B7, B10, B7+3 and B100. Vegetable oil is rarely used as fuel for passenger cars, therefore it was not tested. The driving cycles applied (NEDC and CADC) were both transient and the corresponding soot emissions were sampled together on one filter. The vehicle was equipped with an EGR system and a DPF.

Results are reported in Figure 4-31. The EURO V diesel passenger car showed the highest NPAH emissions of the whole biofuel project with a total NPAH concentration in the exhaust gas of approx. 150 ng/m^3. Despite differences up to 20% between the different fuels tested, no uniform trend could be observed. The most emitted NPAH was by far 1-nitropyrene, which represented more than 90% of the total NPAH emission in all samples. The use of biodiesel or other blends had no effect on its emission. Nevertheless some aspects should be discussed. The first is the artifact 1-nitropyrene formation on the sampling filter. Actually, at the time of the sampling campaign for the biofuels project, no NO_2 denuding system had been applied. Thus, even by sampling from a dilution tunnel (15.8 dilution factor), since the sampling time was relatively long (80 min) artifact NPAH formation cannot be excluded. The second aspect that should be considered is the nature of the aftertreatment system. The EURO V passenger car was equipped with an EGR and a DPF. With such a configuration, due to the absence of any aftertreatment control for NO_x, the engine had to fulfill the NO_x emission limits "alone". This is mainly achieved by the use of the EGR and by retarding the fuel injection. Because of the already cited trade-off effect, under such conditions combustion is not complete, more soot is produced and that may result in enhanced PAH and NPAH formation. Despite of that, 1-nitropyrene could be also newly produced in the DPF.

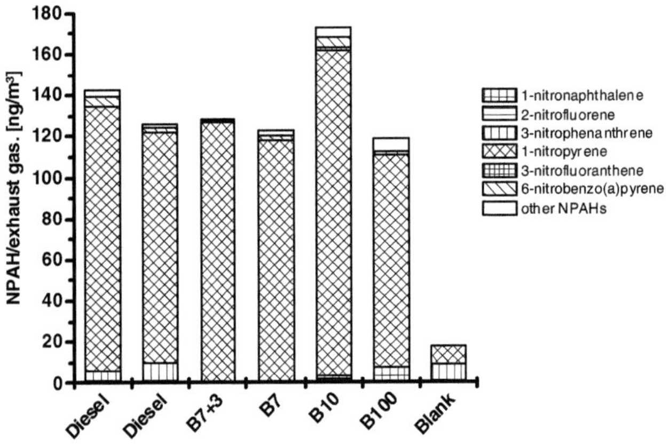

Figure 4-31. NPAH emission from a EURO V passenger car.

The emission profile is indeed rather similar to that of the active regeneration showed in

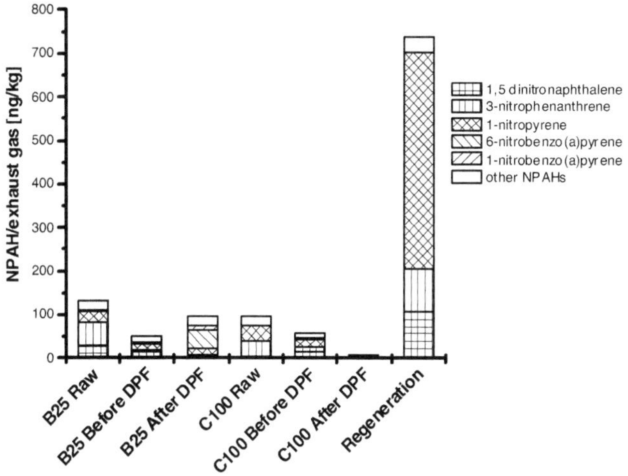

Figure 4-28, where the emission of 1-nitropyrene is dominant. It could be that the 1-nitropyrene originated during the active filter regeneration thus covering any effect of the use of biodiesel on NPAH emission.

4.4.3 EURO IV passenger car (otto-cycle)

At the Technical University of Vienna the NPAH emissions originated from an otto-cycle EURO IV passenger car were analyzed. Emissions of fossil gasoline were compared with those of blends E03, E05, E10, E75 and E85. Driving cycles were again the transient NEDC and CADC resulting in 80 min sampling time.

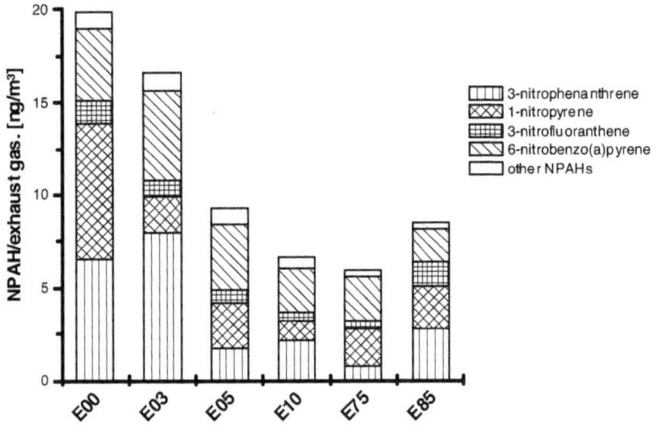

Figure 4-32. NPAH emission from a EURO IV passenger car (otto cycle).

Results are shown in Figure 4-32 and reveal the reduction of the total NPAH concentration in exhaust gas by increasing the ethanol percentage in the fuel. The use of fossil gasoline produced a total NPAH concentration of 20 ng/m^3. Already introducing 3% of ethanol, emissions were 15% lower than those of pure fossil gasoline and they decreased to 6 ng/m^3 by the use of E10. An increase of the ethanol part to 75% and 85% (E75 and E85) did not decrease the NPAH emissions any further. Considering the NPAH emission profile, the most emitted compounds were 3-nitrophenanthrene, 1-nitropyrene, 6-nitrobenzo(a)pyrene and 3-nitrofluoranthene which account for more than 90% of the total emission. The use of ethanol did not affect the emission any NPAH in particular, thus the emission profile remained substantially unchanged with all blends tested.

4.4.4 EU III A-Engine

NPAH emitted from a tractor engine were analyzed too. The engine was at the test bench of the Technical University in Graz. The test cycles used were a transient one (NRTC), which simulates a normal driving behavior and a stationary one (NRSC), where 8 different combinations of load and engine speed are hold for a given time. NRTC cycle was driven twice resulting in a sampling time of 40 min.

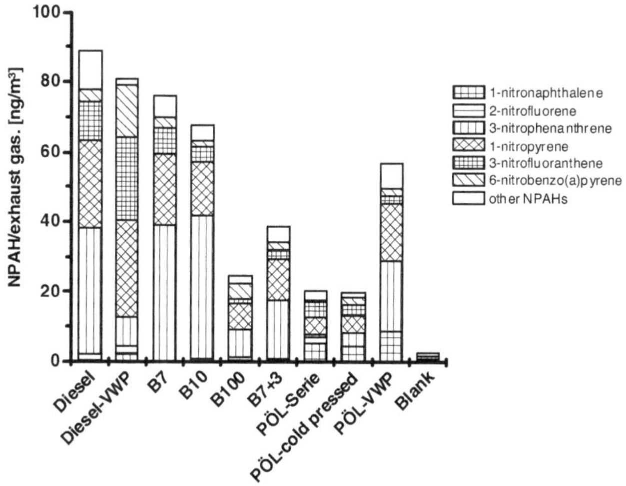

Figure 4-33. NPAH emitted from a tractor engine (NRTC cycle).

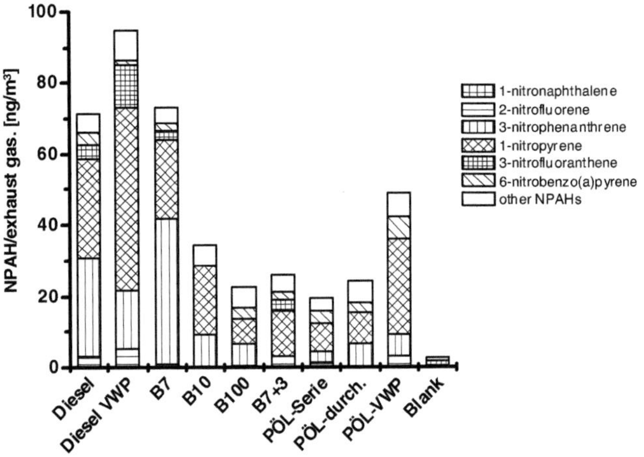

Figure 4-34.NPAH emission from a tractor engine (NRSC cycle).

Sampling time for NRSC cycle was 100, min but the emissions in the transition phase from one last point to the next one were not sampled with exception of one measurement (labeled as PÖL-durch.). The engine was operated with fossil diesel fuel, B100 and vegetable oil (PÖL), and with the blends B7, B10 and B7+3. Vegetable oil was tested both as refined product (serie) and as cold-pressed. In addition, a special injection system (VWP) optimized for vegetable oil was compared with the normal one.

Results regarding NPAH emission during the transient cycle and stationary cycle are shown in Figure 4-33 and Figure 4-34 respectively. Consistently with the PAH analysis, NPAH concentrations in the exhaust gas were similar for both cycles (15). These results show rather clearly that the use of vegetable oil had positive effects on the emissions of NPAH which were reduced from 70 ng/m^3 - 90 ng/m^3 to 20 ng/m^3 - 50 ng/m^3. Surprisingly, the VWP system optimized for the injection of vegetable oil, did not produce any benefit. In contrary, emissions of NPAH were slightly increased. Pure biodiesel (B100) and B7+3 blend, caused significant reduction comprised between 70% and 50% in comparison to fossil diesel fuel both during the transient cycle and during the stationary. The blend B10 also reduced NPAH emission, but the reduction is more pronounced during the stationary cycle.

Considering the NPAH emission profile, 3-nitrophenanthrene, 1-nitropyrene and 3-nitrofluorathene were the most emitted compounds. Their concentration in the exhaust accounted for the 70% - 90% of the total NPAH emission. Emission of nitro-naphthalenes are reduced during the stationary cycle in comparison to the transient one und that is nicely consistent with the analysis of the NPAH emission from the EURO V HD truck see (section 4.4.1).

5 Summary

The present dissertation investigated the interaction of soot-adsorbed PAH with NO_2 both in laboratory model systems and in real diesel engines. Scope of the experiments was to explain to which extent modern DPFs could support the formation of highly toxic and mutagenic NPAHs by nitration of pre-existing PAHs. In addition, the use of biofuels and biofuel blends in different vehicles was investigated with respect to the effects on NPAH emission.

First, an artificial soot aerosol consisting of spark discharge soot particles was reproducibly coated with pyrene and benzo(a)pyrene. The obtained sub-monolayer coating was selected to simulate the adsorption conditions of PAHs on diesel soot. The heterogeneous reaction of adsorbed PYR and BaP with NO_2 was investigated at 25°C for the determination of NPAH formation. Focus of these experiments was the artifact NPAH formation under sampling conditions and the investigation of the nitration reaction in DPFs at low temperature (e.g. cold start or idle).

In regard to PYR-coated soot, 1-NPYR was found as the sole nitration product and its production was linearly correlated with reaction time. After 180 minutes under a NO_2 concentration of 4 ppm, its production accounts for ca. 10% of the initially deposited PYR. Under these conditions the 1-NPYR formation rate was 1.08×10^{-3} $\mu g_{1\text{-NPYR}}$ mg_{soot}^{-1} min^{-1}. A partial reaction order a = 1.52 with respect to NO_2 concentration was calculated. In particular, after 30 min, production of 1-NPYR was noticed already at the lowest investigated concentration of 0.11 ppm NO_2. By increasing the reaction time up to 180 min and the $[NO_2]$ up to 4 ppm, the 1-NPYR production increased by a factor of 1000. On the basis of these results a radical addition elimination mechanism of reaction has been proposed.

BaP has been found to be more reactive than PYR. After 120 minutes under a NO_2 concentration of 5 ppm almost 80% of the BaP was converted into 6-NBaP. These results demonstrate the occurrence of a heterogeneous nitration reaction, which transformed PAH into NPAH like it could occur in DPFs at low temperature during cold start or idling phases (e.g.

at a traffic light or during a jam). Moreover, the need of efficient NO_2 scrubbing systems to avoid NPAH artifacts formation on sampling filters could be confirmed.

Investigating the effects of temperature, we found the highest 1-NPYR concentration on particles at 100°C, whereas at temperature above 200°C the 1-NPYR amount found was comparable to that found at 25°C. Backed by corroborating results from separate gas/solid phase partition experiments with 1-NPYR and 6-NBaP, it is likely that the adsorbed PAH and the newly formed 1-NPYR became transferred from particle to gas phase at higher temperatures. This is of importance for real DPFs operation conditions because at the working temperature of these systems (200°C – 450°C), a significant fraction of the NPAH could be found in the gas phase. Furthermore, experiments with PYR- and BaP-coated soot showed a significant NPAH formation already at 25°C. Therefore, the hypothesis of a heterogeneous NPAH formation in DPFs during the cold-start or idling phase in the temperature range 25°C – 100°C followed by desorption at higher temperature is plausible.

The second series of experiments involved BaP-coated soot and whose nitration reaction was investigated in the DPF simulation system under diesel exhaust relevant specifications. After 1 hour of sampling time and with NO_2 concentration of 120 ppm, production of 1-NBaP and 6-NBaP were observed within the DPF structures in the temperature range 25°C – 250°C. Such residence times and concentrations are comparable with the conditions of DPF regeneration. To find the reaction products by extraction of the soot deposited within the DPF structure demonstrates that the reaction could occur heterogeneously on the soot surface. The highest NPAH production was observed at 150°C and decreased at 250°C. With increasing temperature a significant fraction of BaP was found in the gas phase and the newly formed NPAH desorbed directly into the gas phase. This is of a certain importance for real DPFs system where the high temperature could cause a sublimation of the NPAH which would be emitted in the gas phase. That is even more relevant for PAH which have a lower vapor pressure than NPAH. Therefore, PAHs could sublimate into the gas phase making them unavailable for the nitration on the soot particles. Hence, there is the concrete possibility that the production of 6-NBaP, and NPAH in general, takes place directly in the gas phase. Future experiments should be addressed to investigate this aspect. However, since in our model system the inlet aerosol was NPAH-free, it can be affirmed that the DPF structure supported the nitration chemistry of BaP with the formation of its highly mutagenic nitro-derivatives 1- and 6-nitrobenzo(a)pyrene.

For the completeness of the dissertation, NPAH emissions from a heavy duty engine were analyzed with focus on the effects of the CRT aftertreatment system. In such systems the PAH and NPAH degradation process may compete with the NPAH formation. It was found that the CRT reduced the total amount of PAH and NPAH emitted both at low load and high load. Nevertheless, at low load the lower temperatures and the higher NO_2 concentration furnished better conditions for PAH nitration while they were more adverse to NPAH degradation. These effects were NPAH-specific. For these reasons, some NPAH like 3-nitrophenanthrene were still efficiently degraded but others were newly formed. For instance emissions of the highly toxic 1-benzo(a)pyrene and 6-nitrobenzo(a)pyrene where increased by a approx. a factor 10 at low load and, even if in lower amount, were formed also at high load. The super-mutagen 1,6-dinitropyrene, which was not present in raw exhaust, could be found only after the DPF, indicating a new formation. In regard to emissions from the active regeneration mode, tremendous high emissions of 1-nitropyrene were reported. More detailed investigation of such a regeneration mode would be of great interest for the future

On the basis of the laboratory and the real engine emissions data presented in this work, it can be concluded that PAH nitration in diesel particulate filter systems occurs and can influence the emission pattern of NPAH. Different reaction pathways are proposed. The reaction is to some extent heterogeneous and occurs on the particles surface. With increasing temperature this reaction pathway should become less relevant and a gas-phase reaction could gain importance. In real diesel engines, the nitration reaction is in competition with the degradation of engine-out PAH and NPAH, which is favored at temperature in heavy duty engines > 350°C (high load). Since some PAH (e.g. benzo(a)pyrene) are much more prone to nitration than others and the degradation efficiency is different for every NPAH, with the use of diesel particulate filters a modification of the PAH and NPAH emissions profile from diesel engines can be expected.

The study of NPAH emission from CRT applied to passenger cars would be of great interest. In fact, exhaust gas temperatures of passenger cars are between 100°C and 200°C lower than those of heavy duty vehicles and that could have an effect on the competitive processes of

NPAH degradation vs. NPAH new formation. Therefore, sampling of passenger car emissions should be planned as next step for future investigations. Finally, toxicological studies should be conducted to determine which aftertreatment system is more favorable to humans. Those results should be used to assess risks and benefits of the DPF technology.

In regard to NPAH emissions from different vehicles depending on the use of biofuels, vehicle-specific responses were observed. In fact, emissions form the EURO 5 truck were not influenced by the use of biodiesel, vegetable oil and their blends. NPAH concentrations in exhaust gas were overall rather low and no significant difference between biofuels and fossil diesel fuel could be determined.

The EURO 5 diesel passenger showed the highest NPAH concentrations among all vehicles tested. Emissions were dominated from 1-nitropyrene but emissions derived from usage of biofuels were not significantly different from those of the reference fossil fuel. Despite of that, the role of the DPF which was mounted on the passenger car should be considered. Actually, the NPAH emission profile was similar to that of an active regeneration. Therefore eventual differences in NPAH emissions caused from the use of biofuels could be covered form the NPAH formation in the aftertreatment system.

NPAH emissions from the otto-cycle passenger car, decreased with the increase of the blended ethanol amount. Already with 3% of ethanol, positive effects could be observed, and by the use of 10% of ethanol NPAH emissions decreased by 70% in comparison to fossil gasoline. However, by increasing of the ethanol content up to 75% and 85% (E75 and E85), NPAH emissions were not further reduced.

Finally, positive effects could be observed also by operating a tractor engine with biodiesel and vegetable oil. NPAH emissions were reduced by the use of pure biofuels, but positive effects up to 50% reduction in comparison to fossil diesel could be observed also by blends B10 and B7+3.

In conclusion, it can be affirmed that the use of biofuels did not produce any significant variation in the NPAH emissions in two vehicles (EURO 5 truck and EURO 5 diesel passenger car). On the other hand, by operating the tractor engine and the otto-cycle passenger car with biodiesel and ethanol respectively, the NPAH concentrations in the exhaust were significantly reduced.

6 Appendix

6.1 Abbreviation list

1-NBaP	1-nitrobenzozo(a)pyrene
1-NPYR	1-nitropyrene
2,7-DNFLU	2,7-dinitrofluorene
2-NFLT	2-nitrofluoranthene
2-NPYR	2-nitropyrene
3-NBaP	3-nitrobenzozo(a)pyrene
6-NBaP	6-nitrobenzozo(a)pyrene
ARTEMIS	Assessment and Reliability of Transport Emission Models and Inventory Systems
BaP	Benzo(a)pyrene
CADC	Common ARTEMIS driving cycle
CNC	Condensation nuclei counter
CPC	Condensation particle counter
CVS	Constant volume sampler
DCM	Dichloromethane
DMA	Differential mobility analyzer
DPF	Diesel particulate filter
EC	Elemental carbon
EPA	Environmental protection agency
ESC	European stationary cycle
ETC	European transient cycle
FAME	Fatty acid methyl ester
FBC	Fuel born catalyst
FBR	Flat bed reactor
FIGE	Forschungsinstitut Geräusche und Erschütterungen
GC	Gas chromatography
GfG	Spark discharge soot
HC	Hydrocarbons

HD	Heavy duty
HPLC	High performance liquid chromatography
I.S-.	Internal standard
LNT	Lean NO_x trap
LOE	Limit of emission
MeOH	Methanol
NEDC	New European driving cycle
NPAH	Nitrated polycyclic aromatic hydrocarbons
NRSC	Non road stationary cycle
NRTC	Non road transient cycle
OC	Organic carbon
PAH	Polycyclic aromatic hydrocarbons
PCDD/Fs	Polychlorinated dioxins/furans
PM	Particulate matter
PM1	Particles with aerodynamic diameter < 1 µm
PM10	Particles with aerodynamic diameter < 10 µm
PYR	Pyrene
ROS	Reactive oxygen species
SCR	Selective catalyst reduction
SFG	Surface functional gropus
SiC	Silicon carbide
SMPS	Scanning mobility particle sizer
SOF	Soluble organic fraction
TEM	Transmission electron microscopy

6.2 NPAH structures

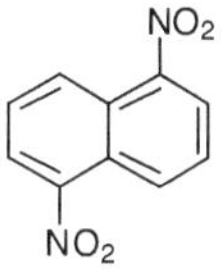

1,5-dinitronaphthalene

1

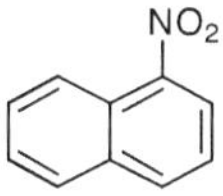

1-nitronaphthalene

2

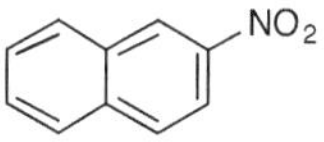

2-nitronaphthalene

3

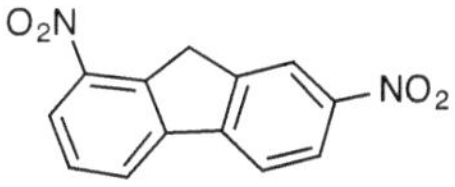

2,7-dinitrofluorene
4

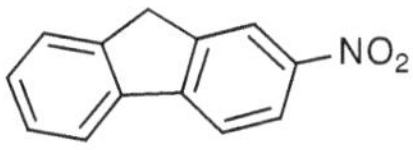

2-nitrofluorene
5

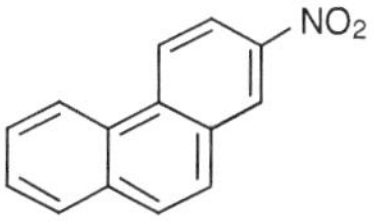

3-nitrophenathrene
6

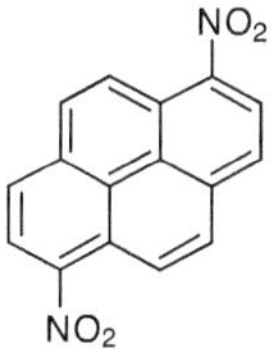

1,6-dinitropyrene
7

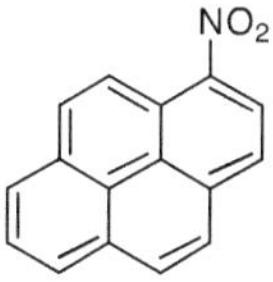

1-nitropyrene
8

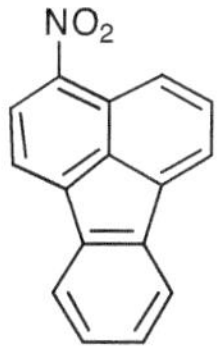

3-nitrofluoranthene
9

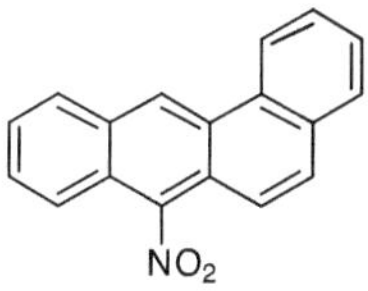

7-nitrobenz(a)anthracene
10

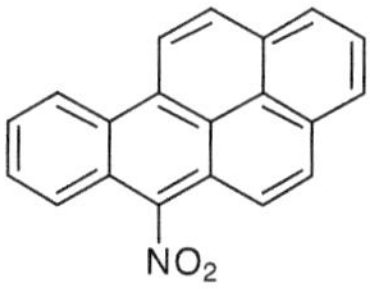

6-nitrobenzo(a)pyrene
11

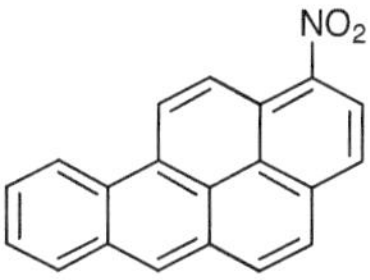

1-nitrobenzo(a)pyrene
12

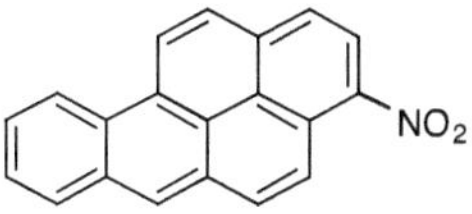

3-nitrobenzo(a)pyrene

13

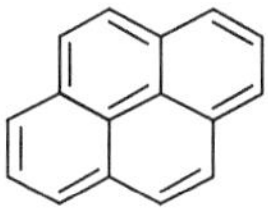

pyrene

14

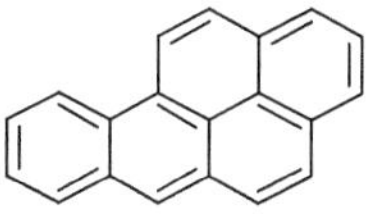

benzo(a)pyrene

15

6.3 Chemicals

Gases

Argon (99.995%, Messer-Griesheim, Germany)

Nitrogen (99.999%, Air liquide, Germany)

Nitrogen dioxide (1.018% in synthetic air, Messer-Griesheim, Germany)

Oxygen (99.998%, Air Liquide, Germany)

Chemicals

1,5-dinitronaphthalene (98%, Aldrich, Germany)

1,6-dinitropyrene 10 mg/L (Dr. Ehrenstorfer, Germany)

1-nitrobenzo(a)pyrene (own synthesis)

1-nitronaphthalene 10 mg/L (Dr. Ehrenstorfer, Germany)

1-nitropyrene (> 99 %, Sigma-Aldrich, Germany)

2,7-dinitrofluorene (> 99 %, Dr. Ehrenstorfer, Germany)

2-nitrofluorene 10 mg/L (Dr. Ehrenstorfer, Germany)

2-nitronaphthalene10 mg/L (Dr. Ehrenstorfer, Germany)

3-nitrobenzo(a)pyrene (own synthesis)

3-nitrofluoranthene 10 mg/L (Dr. Ehrenstorfer, Germany)

3-nitrophenanthrene 10 mg/L (Dr. Ehrenstorfer, Germany)

6-nitrobenzo(a)pyrene 10 mg/L (Dr. Ehrenstorfer, Germany)

7-nitrobenzo(a)anthracene 10 mg/L (Dr. Ehrenstorfer, Germany)

Acetic acid (> 96.5 %, Sigma Aldrich, Germany)

Benzo(a)pyrene (puriss, > 98 %, Sigma-Aldrich, Germany)

Ethylene glycol (> 99.5 %, Fluka, Germany)

Guiacol (> 99 %Sigma, Germany)

N-(1-Naphtyl)etylendiaminedihydrochloride (98%, Aldrich, Germany)

Platinum/γ-aluminum oxide (5% Pt, Merck, Germany))

Potassium hydroxide (> 85 %, Roth, Germany)

Potassium iodide (> 99.5%, Roth, Germany)

Pyrene (99%, Aldrich, Germany)

Silica gel (70-230, Sigma-Aldrich, Germany)

Sodium arsenite (reagent grade, fisher scientific, United Kingdom)

Sodium nitrite (> 99 % Riedel-de Haen, Germany)

Solvents

Acetone (> 99.5 %, Carl Roth GmbH, Karlsruhe, Deutschland)

Acetonitrile (puriss., > 99.5 %, Sigma-Aldrich, Germany)

Cyclohexane (puriss., > 99.5 %, Sigma-Aldrich, Germany)

Dichloromethane (puriss., > 99.5 %, Sigma-Aldrich, Germany)

Ethanol (puriss., > 99.5 %, Sigma-Aldrich, Germany)

Isopropanol (> 99.5 %, Carl Roth GmbH, Germany)

Methanol (puriss., > 99.5 %, Sigma-Aldrich, Germany)

Toluene (puriss., > 99.5 %, Sigma-Aldrich, Germany)

6.4 Instrumentation

Chemiluminescence NO/NO_x Analyzer

CLD 700 AL (Eco Physics, Germany)

HPLC-FLD

HPLC system Promincence LC-20 (Shimadzu, Germany) composed of:

Autosampler SIL-20A HT

Communications bus module CBM-20A

Pumps LC-20AT

Reduction column oven Croco-cil oven

Separation column oven CTO-10AS VP

Diode array detector SPD-M20A (Shimadzu, Germany)

Fluorescence detector RF-10AXL (Shimadzu, Germany)

Reduction column, 50 mm x 4 mm, (Macherey-Nagel, Germany)

Separation column, Pinnacle-II 150 mm x 3.2 mm, 4 µm (Restek, Germany)

Software LabSolution (Shimadzu, Deutschland)

SMPS measurements

Electrostatic classifier, Model 3071 (TSI, Germany)

CPC, Model 3025 (TSI, Germany)

Software, Aerosol Instrument Manager (TSI, Germany)

UV-vis spectrometer

UV-DU650 (Beckman Coulter Inc., Fullerton, CA, USA)

Others materials

Heated magnetic stirrer, RCT basic (IKA Labortechnik, Germany)

Heating rods, firerod 10 mm x 50 mm, 150W, 19.46 W/cm^2(Watlow GmbH, Germany)

Heating tape, HA-010 (Hillesheim, Germany)

Heating unit, PID HT52 (Hillesheim, Germany)

Polyurethane foams, GA3035 (Ziemer Chromatography, Germany)

Precision scale, AT 261 Delta Range (Mettler, Germany)

Quartz fiber filters 47 mm and 70 mm (Advantec, Japan)

Rotating evaporator, Rotavapor (Büchi, Swizerland)

Spark discharge soot generator, GFG 1000 (Palas, Germany)

Temperature sensor, K HKMTSS-150 (Newport omega, Germany)

Tubes, Tygon R 3603 (Roth, Germany)

Ultra-pure water system, MilliQ plus 185 (Millipore, USA)

Ultrasonic bath, Sonorex RK510S (Bandelin, Germany)

7 Literature

(1) Monks, P. S.; Granier, C.; Fuzzi, S.; Stohl, A.; Williams, M. L.; Akimoto, H.; Amann, M.; Baklanov, A.; Baltensperger, U.; Bey, I.; Blake, N.; Blake, R. S.; Carslaw, K.; Cooper, O. R.; Dentener, F.; Fowler, D.; Fragkou, E.; Frost, G. J.; Generoso, S.; Ginoux, P.; Grewe, V.; Guenther, A.; Hansson, H. C.; Henne, S.; Hjorth, J.; Hofzumahaus, A.; Huntrieser, H.; Isaksen, I. S. A.; Jenkin, M. E.; Kaiser, J.; Kanakidou, M.; Klimont, Z.; Kulmala, M.; Laj, P.; Lawrence, M. G.; Lee, J. D.; Liousse, C.; Maione, M.; McFiggans, G.; Metzger, A.; Mieville, A.; Moussiopoulos, N.; Orlando, J. J.; O'Dowd, C. D.; Palmer, P. I.; Parrish, D. D.; Petzold, A.; Platt, U.; Poeschl, U.; Prevot, A. S. H.; Reeves, C. E.; Reimann, S.; Rudich, Y.; Sellegri, K.; Steinbrecher, R.; Simpson, D.; ten Brink, H.; Theloke, J.; van der Werf, G. R.; Vautard, R.; Vestreng, V.; Vlachokostas, C.; von Glasow, R. Atmospheric composition change - global and regional air quality. *Atmospheric Environment.* **2009**, *43*, 5268-5350.

(2) Poeschl, U. Atmospheric aerosols: composition, transformation, climate and health effects. . *Angewandte Chemie, International Edition.* **2005**, 7520-7540.

(3) Prather, K. A.; Hatch, C. D.; Grassian, V. H. Analysis of atmospheric aerosols. *Annu. Rev. Anal. Chem.* **2008**, *1*, 485-514.

(4) Denman, K. L.; Brasseur, G.; Chidthaisong, A.; Ciais, P.; Cox, P. M.; Dickinson, R. E.; Hauglustaine, D.; Heinze, C.; Holland, E.; Jacob, D.; Lohmann, U.; Ramachandran, S.; Dias, P. L. d. S.; Wofsy, S. C.; X. Zhang *Couplings between changes in the climate system and biogeochemistry* Climate Change 2007: The Physical Science Basis. Contribution of Working Group I to the Fourth Assessment. Cambridge University Press, Cambridge, United Kingdom and New York, NY, USA., 2007.

(5) Ravishankara, A. R. Heterogeneous and multiphase chemistry in the troposphere. *Science.* **1997**, *276*, 1058-1065.

(6) Avino, P.; Brocco, D. Carbonaceous aerosol in the breathable particulate matter (PM10) in urban area. *Annali di Chimica (Rome, Italy).* **2004**, *94*, 647-653.

(7) Winiwarter, W.; Kuhlbusch, T. A. J.; Viana, M.; Hitzenberger, R. Quality considerations of European PM emission inventories. *Atmospheric Environment.* **2009**, *43*, 3819-3828.

(8) Charron, A.; Harrison, R. M. Primary particle formation from vehicle emissions during exhaust dilution in the roadside atmosphere. *Atmospheric Environment.* **2003**, *37*, 4109-4119.

(9) Lepore, L.; Brocco, D.; Avino, P. Organic and elemental carbon in atmospheric particles. *Ann Ist Super Sanita FIELD Full Journal Title:Annali dell'Istituto superiore di sanita.* **2003**, *39*, 365-369.

(10) European council directive 1999/30/EC Sulphur dioxide, nitrogen dioxide and oxides of nitrogen, particulate matter and lead in ambient air 1999

(11) Perez, N.; Pey, J.; Cusack, M.; Reche, C.; Querol, X.; Alastuey, A.; Viana, M. Variability of particle number, black carbon, and PM10, PM2.5, and PM1 levels and speciation: influence of road traffic emissions on urban air quality. *Aerosol Science and Technology.* **2010**, *44*, 487-499.

(12) Keywood, M. D.; Ayers, G. P.; Gras, J. L.; Gillett, R. W.; Cohen, D. D. Relationships between size segregated mass concentration data and ultrafine particle number concentrations in urban areas. *Atmospheric Environment.* **1999**, *33*, 2907-2913.

(13) Tuch, T.; Brand, P.; Wichmann, H. E.; Heyder, J. Variation of particle number and mass concentration in various size ranges of ambient aerosols in Eastern Germany. *Atmospheric Environment.* **1997**, *31*, 4193-4197.

(14) Kreyling, W. G.; Tuch, T.; Peters, A.; Pitz, M.; Heinrich, J.; Stolzel, M.; Cyrys, J.; Heyder, J.; Wichmann, H. E. Diverging long-term trends in ambient urban particle mass and number concentrations associated with emission changes caused by the German unification. *Atmospheric Environment.* **2003**, *37*, 3841-3848.

(15) Knauer, M. *Struktur-Reaktivitäts-Korrelation von Dieselruß und Charakterisierung von PAHs und Carbonylen im Abgas von Biokraftstoffen*; PhD thesis, Technische Universität München, 2009.

(16) *Monographs on the elvaluation of carcinogenic risk on humans*. International Agency for Research on Cancer (IARC), 2008.

(17) *Health assessment document for diesel engine exhaust*; Prepared by the National Center for Environmental Assesment, Washinton, DC, for the office of Transportation and Air Quality , EPA /6008-90/057F; U.S Environmental Protection Agency, National Technical Information Service: Springfield, VA, 2002.

(18) Xi, J.; Zhong, B.-J. Soot in diesel combustion systems. *Chemical Engineering & Technology.* **2006**, *29*, 665-673.

(19) Von Klot, S.; Peters, A.; Aalto, P.; Bellander, T.; Berglind, N.; D'Ippoliti, D.; Elosua, R.; Hoermann, A.; Kulmala, M.; Lanki, T.; Loewel, H.; Pekkanen, J.; Picciotto, S.; Sunyer, J.; Forastiere, F. Ambient air pollution is associated with increased risk of hospital cardiac readmissions of myocardial infarction survivors in five European cities. *Circulation.* **2005**, *112*, 3073-3079.

(20) Lanki, T.; Pekkanen, J.; Aalto, P.; Elosua, R.; Berglind, N.; D'Ippoliti, D.; Kulmala, M.; Nyberg, F.; Peters, A.; Picciotto, S.; Salomaa, V.; Sunyer, J.; Tiittanen, P.; von Klot, S.; Forastiere, F. Associations of traffic related air pollutants with hospitalisation for first acute myocardial infarction: the HEAPSS study. *Occupational and Environmental Medicine.* **2006**, *63*, 844-851.

(21) Araujo, J. A.; Nel, A. E. Particulate matter and atherosclerosis: role of particle size, composition and oxidative stress. *Part. Fibre Toxicol.* **2009**, *6*.

(22) Oberdorster, G. Toxicology of ultrafine particles: in vivo studies. *Philosophical Transactions of the Royal Society of London Series a-Mathematical Physical and Engineering Sciences.* **2000**, *358*, 2719-2739.

(23) El Haddad, I.; Marchand, N.; Dron, J.; Temime-Roussel, B.; Quivet, E.; Wortham, H.; Jaffrezo, J. L.; Baduel, C.; Voisin, D.; Besombes, J. L.; Gille, G. Comprehensive primary particulate organic characterization of vehicular exhaust emissions in France. *Atmospheric Environment.* **2009**, *43*, 6190-6198.

(24) Gualtieri, M.; Mantecca, P.; Corvaja, V.; Longhin, E.; Perrone, M. G.; Bolzacchini, E.; Camatini, M. Winter fine particulate matter from Milan induces morphological and functional alterations in human pulmonary epithelial cells (A549). *Toxicol. Lett.* **2009**, *188*, 52-62.

(25) BeruBe, K.; Balharry, D.; Sexton, K.; Koshy, L.; Jones, T. Combustion-derived nanoparticles: mechanisms of pulmonary toxicity. *Clin. Exp. Pharmacol. Physiol.* **2007**, *34*, 1044-1050.

(26) Ma, J. Y. C.; Ma, J. K. H. The dual effect of the particulate and organic components of diesel exhaust particles on the alteration of pulmonary immune/inflammatory responses and metabolic enzymes. *J. Environ. Sci. Health Pt. C-Environ. Carcinog. Ecotoxicol. Rev.* **2002**, *20*, 117-147.

(27) Liu, Z. G.; Berg, D. R.; Swor, T. A.; Schauer, J. J. Comparative analysis on the effects of diesel particulate filter and selective catalytic reduction systems on a wide spectrum of chemical species emissions. *Environmental Science & Technology.* **2008**, *42*, 6080-6085.

(28) Alkemade, U. G.; Schumann, B. Engines and exhaust after treatment systems for future automotive applications. *Solid State Ionics.* **2006**, *177*, 2291-2296.

(29) van Setten, B.; Makkee, M.; Moulijn, J. A. Science and technology of catalytic diesel particulate filters. *Catalysis Reviews-Science and Engineering.* **2001**, *43*, 489-564.

(30) Rhead, M. M.; Hardy, S. A. The sources of polycyclic aromatic compounds in diesel engine emissions. *Fuel.* **2003**, *82*, 385-393.

(31) Kubatova, A.; Steckler, T. S.; Gallagher, J. R.; Hawthorne, S. B.; Picklo, M. J. Toxicity of wide-range polarity fractions from wood smoke and diesel exhaust particulate obtained using hot pressurized water. *Environmental Toxicology and Chemistry.* **2004**, *23*, 2243-2250.

(32) Reyes, D. R.; Rosario, O.; Rodriguez, J. F.; Jimenez, B. D. Toxic evaluation of organic extracts from airborne particulate matter in Puerto Rico. *Environmental Health Perspectives.* **2000**, *108*, 635-640.

(33) De Martinis, B. S.; Kado, N. Y.; de Carvalho, L. R. F.; Okamoto, R. A.; Gundel, L. A. Genotoxicity of fractionated organic material in airborne particles from Sao Paulo, Brazil. *Mutat. Res. Genet. Toxicol. Environ. Mutagen.* **1999**, *446*, 83-94.

(34) Marquardt, H.; Schäfer, S. *Lehrbuch der Toxikologie*; Wissenschaftliche Verlagsgesellschaft GmbH: Stuttgart, Germany, 2004.

(35) Heeb, N. V.; Schmid, P.; Kohler, M.; Gujer, E.; Zennegg, M.; Wenger, D.; Wichser, A.; Ulrich, A.; Gfeller, U.; Honegger, P.; Zeyer, K.; Emmenegger, L.; Petermann, J.-L.; Czerwinski, J.; Mosimann, T.; Kasper, M.; Mayer, A. Impact of low- and high-oxidation Diesel Particulate Filters on genotoxic exhaust constituents. *Environmental Science & Technology.* **2010**, *44*, 1078-1084.

(36) Esteve, W.; Budzinski, H.; Villenave, E. Relative rate constants for the heterogeneous reactions of NO_2 and OH radicals with polycyclic aromatic hydrocarbons adsorbed on carbonaceous particles. Part 2: PAHs adsorbed on diesel particulate exhaust SRM 1650a. *Atmospheric Environment.* **2006**, *40*, 201-211.

(37) Guo, Z.; Kamens, R. M. An experimental technique for studying heterogeneous reactions of polyaromatic hydrocarbons on particle surfaces. *Journal of Atmospheric Chemistry.* **1991**, *12*, 137-151.

(38) Heeb, N. V.; Schmid, P.; Kohler, M.; Gujer, E.; Zennegg, M.; Wenger, D.; Wichser, A.; Ulrich, A.; Gfeller, U.; Honegger, P.; Zeyer, K.; Emmenegger, L.; Petermann, J.-L.; Czerwinski, J.; Mosimann, T.; Kasper, M.; Mayer, A. Secondary effects of catalytic Diesel particulate filters: conversion of PAHs versus formation of Nitro-PAHs. *Environmental Science & Technology.* **2008**, *42*, 3773-3779.

(39) Hartung, A.; Kraft, J.; Schulze, J.; Kiess, H.; Lies, K. H. The identification of nitrated polycyclic aromatic-hydrocarbons in diesel particulate extracts and their potential formation as artifacts during particulate collection. *Chromatographia.* **1984**, *19*, 269-273.

(40) Levsen, K.; Puttins, U.; Schilhabel, J.; Priess, B. Artifacts during the sampling of nitro-PAHs of diesel exhaust. *Fresenius Zeitschrift Fur Analytische Chemie.* **1988**, *330*, 527-528.

(41) Poeschl, U.; Letzel, T.; Schauer, C.; Niessner, R. Interaction of ozone and water vapor with spark discharge soot aerosol particles coated with benzo[a]pyrene: O_3 and H_2O adsorption, benzo[a]pyrene degradation, and atmospheric implications. *Journal of Physical Chemistry* **2001**, *105*, 4029-4041.

(42) Wallington, T. J.; Kaiser, E. W.; Farrell, J. T. Automotive fuels and internal combustion engines: a chemical perspective. *Chem. Soc. Rev.* **2006**, *35*, 335-347.

(43) Neeft, J. P. A.; Makkee, M.; Moulijn, J. A. Diesel particulate emission control. *Fuel Process. Technol.* **1996**, *47*, 1-69.

(44) Rothe, D. *Physikalische und chemische Charakterisierung der Rußpartikelemission von Nutzfahrzeugdieselmotoren und Methoden zur Emissionsminderung*; PhD thesis, Technische Universität München, 2006.

(45) Arens, F.; Gutzwiller, L.; Baltensperger, U.; Gaggeler, H. W.; Ammann, M. Heterogeneous reaction of NO_2 on diesel soot particles. *Environmental Science & Technology.* **2001**, *35*, 2191-2199.

(46) Lackner, M.; Winter, F.; Geringer, B. Chemistry in the motor. *Chem. unserer Zeit.* **2005**, *39*, 246-254.

(47) Stanislaus, A.; Marafi, A.; Rana, M. S. Recent advances in the science and technology of ultra low sulfur diesel (ULSD) production. *Catalysis Today.* **2010**, *153*, 1-68.

(48) Zabetta, E. C.; Kilpinen, P. Improved NO_x submodel for in-cylinder CFD simulation of low- and medium-speed compression ignition engines. *Energy Fuels.* **2001**, *15*, 1425-1433.

(49) Jimenez, J. L.; McRae, G. J.; Nelson, D. D.; Zahniser, M. S.; Kolb, C. E. Remote sensing of NO and NO_2 emissions from heavy-duty diesel trucks using tunable diode lasers. *Environmental Science & Technology.* **2000**, *34*, 2380-2387.

(50) Heywood, J. B. *Internal Combustion Engine Fundamentals*; McGraw-Hill, 1988.

(51) Goldsworthy, L. Reduced kinetics schemes for oxides of nitrogen emissions from a slow-speed marine diesel engine. *Energy Fuels.* **2003**, *17*, 450-456.

(52) Fenimore, C. P. Reactions of fuel-nitrogen in rich flame gases. *Combust. Flame.* **1976**, *26*, 249-256.

(53) Zervas, E.; Bikas, G. Impact of the driving cycle on the NO_x and particulate matter exhaust emissions of diesel passenger cars. *Energy Fuels.* **2008**, *22*, 1707-1713.

(54) Chellam, S.; Kulkarni, P.; Fraser, M. P. Emissions of organic compounds and trace metals in fine particulate matter from motor vehicles: A tunnel study in Houston, Texas. *J. Air Waste Manage. Assoc.* **2005**, *55*, 60-72.

(55) Montero, L.; Duane, M.; Manfredi, U.; Astorga, C.; Martini, G.; Carriero, M.; Krasenbrink, A.; Larsen, B. R. Hydrocarbon emission fingerprints from contemporary vehicle/engine technologies with conventional and new fuels. *Atmospheric Environment.* **2010**, *44*, 2167-2175.

(56) Payri, F.; Bermudez, V. R.; Tormos, B.; Linares, W. G. Hydrocarbon emissions speciation in diesel and biodiesel exhausts. *Atmospheric Environment.* **2009**, *43*, 1273-1279.

(57) Xi, J.; Zhong, B. J. Soot in diesel combustion systems. *Chemical Engineering & Technology.* **2006**, *29*, 665-673.

(58) Tree, D. R.; Svensson, K. I. Soot processes in compression ignition engines. *Prog. Energy Combust. Sci.* **2007**, *33*, 272-309.

(59) Smith, O. I. Fundamentals of soot formation in flames with application to diesel engine particulate emissions. *Prog. Energy Combust. Sci.* **1981**, *7*, 275-291.

(60) Krestinin, A. V.; Kislov, M. B.; Raevskii, A. V.; Kolesova, O. I.; Stesik, L. N. On the mechanism of soot particle formation. *Kinet. Catal.* **2000**, *41*, 90-98.

(61) Ishiguro, T.; Takatori, Y.; Akihama, K. Microstructure of diesel soot particles probed by electron microscopy: First observation of inner core and outer shell. *Combust. Flame.* **1997**, *108*, 231-234.

(62) Stanmore, B. R.; Brilhac, J. F.; Gilot, P. The oxidation of soot: a review of experiments, mechanisms and models. *Carbon.* **2001**, *39*, 2247-2268.

(63) Sadezky, A.; Muckenhuber, H.; Grothe, H.; Niessner, R.; Pöschl, U. Raman spectra of soot and related carbonaceous materials: Spectral analysis and structural information. *Carbon.* **2005**, *43*, 1731-1742.

(64) Sze, S.-K.; Siddique, N.; Sloan, J. J.; Escribano, R. Raman spectroscopic characterization of carbonaceous aerosols. *Atmospheric Environment.* **2001**, *35*, 561-568.

(65) Burtscher, H. Physical characterization of particulate emissions from diesel engines: a review. *Journal of Aerosol Science.* **2005**, *36*, 896-932.

(66) Haralampous, O.; Koltsakis, G. C. Intra-layer temperature gradients during regeneration of diesel particulate filters. *Chemical Engineering Science.* **2002**, *57*, 2345-2355.

(67) Mathis, U.; Mohr, M.; Kaegi, R.; Bertola, A.; Boulouchos, K. Influence of diesel engine combustion parametes on primary soot particle diameter. *Environmental Science & Technology.* **2005**, *39*, 1887-1892.

(68) Hasegawa, K.; Kaneko, H.; Ogawa, T. Autocatalytic nitration of pyrene by aerated nitrogen dioxide in solution and comparison with the nitration on silica particles. *Bull. Chem. Soc. Jpn.* . **2004**, *77*, 147-155.

(69) Vander Wal, R. L.; Bryg, V. M.; Hays, M. D. Fingerprinting soot (towards source identification): Physical structure and chemical composition. *Journal of Aerosol Science.* **2010**, *41*, 108-117.

(70) Wal, R. L. V.; Tomasek, A. J.; Street, K.; Thompson, W. K.; Hull, D. R. *Carbon nanostructure examined by lattice fringe analysis of high resolution transmission electron microscopy images.* NASA Center for Aerospace Information, 2003.

(71) Schloegl, R. Surface composition and structure of active carbons. *Handbook of Porous Solids.* **2002**, *3*, 1863-1900.

(72) Kittelson, D. B. Engines and nanoparticles: A review. *Journal of Aerosol Science.* **1998**, *29*, 575-588.

(73) Fraser, M. P.; Cass, G. R.; Simoneit, B. R. T. Particulate organic compounds emitted from motor vehicle exhaust and in the urban atmosphere. *Atmospheric Environment.* **1999**, *33*, 2715-2724.

(74) Maricq, M. M. Chemical characterization of particulate emissions from diesel engines: A review. *Journal of Aerosol Science.* **2007**, *38*, 1079-1118.

(75) Kim, C.-G. *A crank angle resolved cidi engine combustion model with arbitrary fuel innjection for control purpose*; PhD disseretation, Ohio State University, 2004.

(76) Schmid, M.; Leipertz, A. An approach for the control of combustion and pollutant formation in diesel engines by combined in-cylinder and exhaust gas measurements. *Int. J. Veh. Des.* **2006**, *41*, 188-205.

(77) Lloyd, A. C.; Cackette, T. A. Diesel engines: environmental impact and control. *J. Air Waste Manage. Assoc.* **2001**, *51*, 809-847.

(78) Johnson, T. V. Review of diesel emissions and control. *Int. J. Engine Res.* **2009**, *10*, 275-285.

(79) Parks, J. E. Less costly catalysts for controlling engine emissions. *Science.* **2010**, *327*, 1584-1585.

(80) Konstandopoulos, A. G.; Papaioannou, E. Update on the science and technology of diesel particulate filters. *Kona.* **2008**, *26*, 36-65.

(81) Solvat, O.; Marez, P.; Belot, G. Passenger car serial application of a particulate filter system on a common-rail, direct-injection Diesel engine. *SAE paper 2000-01-0473.* **2000**.

(82) Pallavkar, S.; Kim, T. H.; Rutman, D.; Lin, J.; Ho, T. Active regeneration of diesel particulate filter employing microwave heating. *Industrial & Engineering Chemistry Research.* **2009**, *48*, 69-79.

(83) Adler, J. Ceramic diesel particulate filters. *International Journal of Applied Ceramic Technology.* **2005**, *2*, 429-439.

(84) Brunner, N.; Bloom, R.; Schroeer, S. Fiber-wound diesel particulate filter durability experience with metal-based additives. *SAE paper 970180.* **1997**.

(85) Liati, A.; Eggenschwiler, P. D. Characterization of particulate matter deposited in diesel particulate filters: visual and analytical approach in macro-, micro- and nano-scales. *Combust. Flame.* **2010**, *157*, 1658-1670.

(86) Hanamura, K.; Karin, P.; Cui, L.; Rubio, P.; Tsuruta, T.; Tanaka, T.; Suzuki, T. Micro- and macroscopic visualization of particulate matter trapping and regeneration processes in wall-flow diesel particulate filters. *Int. J. Engine Res.* **2009**, *10*, 305-321.

(87) *Diesel filter materials*; DieselNet Technology Guide, www.dieselNet.com, 2003.

(88) Fino, D.; Specchia, V. Open issues in oxidative catalysis for diesel particulate abatement. *Powder Technol.* **2008**, *180*, 64-73.

(89) Ciambelli, P.; Palma, V.; Russo, P.; Vaccaro, S. Performances of a catalytic foam trap for soot abatement. *Catalysis Today.* **2002**, *75*, 471-478.

(90) Meyer, A.; Egli, H.; Burtscher, H.; Czerwinski, J.; Gehrig, D. Particle size distribution downstream traps of different design. *SAE paper 950373.* **1995**.

(91) MAN AG, offical website www.man.com.

(92) Shyam, A.; Lara-Curzio, E.; Watkins, T. R.; Parten, R. J. Mechanical characterization of diesel particulate filter substrates. *J. Am. Ceram. Soc.* **2008**, *91*, 1995-2001.

(93) Kandylas, I. P.; Haralampous, O. A.; Koltsakis, G. C. Diesel soot oxidation with NO_2: engine experiments and simulations. *Industrial & Engineering Chemistry Research.* **2002**, *41*, 5372-5384.

(94) Shrivastava, M.; Nguyen, A.; Zheng, Z. Q.; Wu, H. W.; Jung, H. S. Kinetics of soot oxidation by NO_2. *Environmental Science & Technology.* **2010**, *44*, 4796-4801.

(95) Hawker, P.; Myers, N.; Huethwohl, G.; Vogel, H.; Bates, B.; Magnusson, L.; Bronnenberg, P. Experience with a new particulate trap technology in europe. *SAE paper 970182.* **1997**.

(96) Carslaw, D. C. Evidence of an increasing NO_2/NO_X emissions ratio from road traffic emissions. *Atmospheric Environment.* **2005**, *39*, 4793-4802.

(97) Stanmore, B. R.; Tschamber, V.; Brilhac, J. F. Oxidation of carbon by NO_x, with particular reference to NO_2 and N_2O. *Fuel.* **2008**, *87*, 131-146.

(98) Jeguirim, M.; Tschamber, V.; Brilhac, J. F. Kinetics of catalyzed and non-catalyzed soot oxidation with nitrogen dioxide under regeneration particle trap conditions. *J. Chem. Technol. Biotechnol.* **2009**, *84*, 770-776.

(99) Jeguirim, M.; Tschamber, V.; Ehrburger, P. Catalytic effect of platinum on the kinetics of carbon oxidation by NO_2 and O_2. *Appl. Catal. B-Environ.* **2007**, *76*, 235-240.

(100) Schejbal, M.; Stepanek, J.; Marek, M.; Koci, P.; Kubicek, M. Modelling of soot oxidation by NO_2 in various types of diesel particulate filters. *Fuel.* **2010**, *89*, 2365-2375.

(101) Maier, N.; Nickel, K. G.; Engel, C.; Mattern, A. Mechanisms and orientation dependence of the corrosion of single crystal Cordierite by model diesel particulate ashes. *J. Eur. Ceram. Soc.* **2010**, *30*, 1629-1640.

(102) Messerer, A.; Niessner, R.; Poeschl, U. Comprehensive kinetic characterization of the oxidation and gasification of model and real diesel soot by nitrogen oxides and oxygen under engine exhaust conditions: Measurement, Langmuir-Hinshelwood, and Arrhenius parameters. *Carbon.* **2005**, *44*, 307-324.

(103) Knauer, M.; Schuster, M. E.; Su, D. S.; Schlögl, R.; Niessner, R.; Ivleva, N. P. Soot structure and reactivity analysis by raman microspectroscopy, temperature-programmed oxidation, and high-resolution transmission electron microscopy. *Journal of Physical Chemistry A.* **2009**, *113*, 13871-13880.

(104) Ivleva, N. P.; Messerer, A.; Yang, X.; Niessner, R.; Pöschl, U. Raman microspectroscopic analysis of changes in the chemical structure and reactivity of soot in a diesel exhaust aftertreatment model system. *Environmental Science & Technology.* **2007**, *41*, 3702-3707.

(105) Knauer, M.; Carrara, M.; Rothe, D.; Niessner, R.; Ivleva, N. P. Changes in structure and reactivity of soot during oxidation and gasification by oxygen, studied by micro-Raman spectroscopy and temperature-programmed oxidation. *Aerosol Science and Technology.* **2009**, *43*, 1-8.

(106) Wenger, D.; Gerecke, A. C.; Heeb, N. V.; Zennegg, M.; Kohler, M.; Naegeli, H.; Zenobi, R. Secondary effects of catalytic diesel particulate filters: Reduced aryl hydrocarbon receptor-mediated activity of the exhaust. *Environmental Science & Technology.* **2008**, *42*, 2992-2998.

(107) Cauda, E.; Fino, D.; Saracco, G.; Specchia, V. Secondary nanoparticle emissions during diesel particulate trap regeneration. *Top. Catal.* **2007**, *42-43*, 253-257.

(108) Bikas, G.; Zervas, E. Regulated and non-regulated pollutants emitted during the regeneration of a diesel particulate filter. *Energy Fuels.* **2007**, *21*, 1543-1547.

(109) Dwyer, H.; Ayala, A.; Zhang, S.; Collins, J.; Huai, T.; Herner, J.; Chau, W. Emissions from a diesel car during regeneration of an active diesel particulate filter. *Journal of Aerosol Science.* **2010**, *41*, 541-552.

(110) Mohr, M.; Forss, A. M.; Lehmann, U. Particle emissions from diesel passenger cars equipped with a particle trap in comparison to other technologies. *Environmental Science & Technology.* **2006**, *40*, 2375-2383.

(111) Heeb, N. V.; Zennegg, M.; Gujer, E.; Honegger, P.; Zeyer, K.; Gfeller, U.; Wichser, A.; Kohler, M.; Schmid, P.; Emmenegger, L.; Ulrich, A.; Wenger, D.; Petermann, J. L.; Czerwinski, J.; Mosimann, T.; Kasper, M.; Mayer, A. Secondarv effects of catalytic diesel particulate filters: Copper-induced formation of PCDD/Fs. *Environmental Science & Technology.* **2007**, *41*, 5789-5794.

(112) Schauer, C.; Niessner, R.; Poeschl, U. Analysis of nitrated polycyclic aromatic hydrocarbons by liquid chromatography with fluorescence and mass spectrometry detection: air particulate matter, soot, and reaction product studies. *Analytical and Bioanalytical Chemistry.* **2004**, *378*, 725-736.

(113) Greenberg, A.; Lwo, J. H.; Atherholt, T. B.; Rosen, R.; Hartman, T.; Butler, J.; Louis, J. Bioassay-directed fractionation of organic-compounds associated with airborne particulate matter - an interseasonal study. *Atmospheric Environment* **1993**, *27*, 1609-1626.

(114) Nielsen, T. Reactivity of polycyclic aromatic hydrocarbons towards nitrating species. *Environmental Science and Technology.* **1984**, *18*, 157-163.

(115) Veigl, E.; Posch, W.; Lindner, W.; Tritthart, P. Selective and sensitive analysis of 1-nitropyrene in diesel exhaust particulate extract by multidimensional HPLC. *Chromatographia.* **1994**, *38*, 199-206.

(116) Vione, D.; Barra, S.; De Gennaro, G.; De Rienzo, M.; Gilardoni, S.; Perrone, M. G.; Pozzoli, L. Polycyclic aromatic hydrocarbons in the atmosphere: monitoring, sources, sinks and fate. II: Sinks and fate. *Ann. Chim.* **2004**, *94*, 257-268.

(117) Reisen, F.; Arey, J. Atmospheric reactions influence seasonal PAH and nitro-PAH concentrations in the Los Angeles basin. *Environmental Science & Technology.* **2005**, *39*, 64-73.

(118) Bamford, H. A.; Bezabeh, D. Z.; Schantz, M. M.; Wise, S. A.; Baker, J. E. Determination and comparison of nitrated-polycyclic aromatic hydrocarbons measured in air and diesel particulate reference materials. *Chemosphere.* **2003**, *50*, 575-587.

(119) Zielinska, B.; Samy, S. Analysis of nitrated polycyclic aromatic hydrocarbons. *Analytical and Bioanalytical Chemistry.* **2006**, *386*, 883-890.

(120) Hayakawa, K.; Murahashi, T.; Akutsu, K.; Kanda, T.; Tang, N.; Kakimoto, H.; Toriba, A.; Kizu, R. Comparison of polycyclic aromatic hydrocarbons and nitropolycyclic aromatic hydrocarbons in airborne and automobile exhaust particulates. *Polycyclic Aromatic Compounds.* **2000**, *20*, 179-190.

(121) Miller-Schulze, J. P.; Paulsen, M.; Toriba, A.; Tang, N.; Hayakawa, K.; Tamura, K.; Dong, L. J.; Zhang, X. M.; Simpson, C. D. Exposures to particulate air pollution and nitro-polycyclic aromatic hydrocarbons among taxi drivers in Shenyang, China. *Environmental Science & Technology.* **2010**, *44*, 216-221.

(122) Kawanaka, Y.; Matsumoto, E.; Wang, N.; Yun, S. J.; Sakamoto, K. Contribution of nitrated polycyclic aromatic hydrocarbons to the mutagenicity of ultrafine particles in the roadside atmosphere. *Atmospheric Environment.* **2008**, *42*, 7423-7428.

(123) Umbuzeiro, G. A.; Franco, A.; Martins, M. H.; Kummrow, F.; Carvalho, L.; Schmeiser, H. H.; Leykauf, J.; Stiborova, M.; Claxton, L. D. Mutagenicity and DNA adduct formation of PAH, nitro-PAH, and oxy-PAH fractions of atmospheric particulate matter from Sao Paulo, Brazil. *Mutat. Res. Genet. Toxicol. Environ. Mutagen.* **2008**, *652*, 72-80.

(124) Bacolod, M. D.; Basu, A. K. Mutagenicity of a single 1-nitropyrene-DNA adduct N-(deoxyguanosin-8-yl)-1-aminopyrene in Escherichia coli located in a GGC sequence. *Mutagenesis.* **2001**, *16*, 461-465.

(125) Ross, D. S.; Hum, G. P.; Schmitt, R. J. Diesel exhaust and pyrene nitration. *Environmental Science and Technology.* **1987**, *21*, 1130-1131.

(126) Landvik, N. E.; Gorria, M.; Arlt, V. M.; Asare, N.; Solhaug, A.; Lagadic-Gossmann, D.; Holme, J. A. Effects of nitrated-polycyclic aromatic hydrocarbons and diesel exhaust particle extracts on cell signalling related to apoptosis: Possible implications for their mutagenic and carcinogenic effects. *Toxicology.* **2007**, *231*, 159-174.

(127) Portet-Koltalo, F.; Oukebdane, K.; Dionnet, F.; Desbene, P. L. Optimization of the extraction of polycyclic aromatic hydrocarbons and their nitrated derivatives from diesel particulate matter using microwave-assisted extraction. *Analytical and Bioanalytical Chemistry.* **2008**, *390*, 389-398.

(128) Perrini, G.; Tomasello, M.; Librando, V.; Minniti, Z. Nitrated polycyclic aromatic hydrocarbons in the environment: Formation, occurrences and analysis. *Ann. Chim.* **2005**, *95*, 567-577.

(129) Schauer, C. *Analyse und Reaktivität von Polyzyklischen Aromatischen Verbindungen in Aerosolen*; PhD thesis, Technische Universität München, 2004.

(130) Fu, P. P.; Herrenosaenz, D.; Vontungeln, L. S.; Lay, J. O.; Wu, Y. S.; Lai, J. S.; Evans, F. E. DNA-adducts and carcinogenicity of nitro-polycyclic aromatic-hydrocarbons. *Environmental Health Perspectives.* **1994**, *102*, 177-183.

(131) Fu, P. P.; Zhan, D. J.; vonTungeln, L. S.; Yi, P.; Qui, F. Y.; HerrenoSaenz, D.; Lewtas, J. Comparative formation of DNA adducts of nitro-polycyclic aromatic hydrocarbons in

mouse and rat liver microsomes and cytosols. *Polycyclic Aromatic Compounds.* **1996**, *10*, 187-194.

(132) Wolf, J. C. *Benzo(a)pyren-Analytik in der Abgasmesstechnik*; Master-Thesis, Technische Universität München, 2009.

(133) Cvacka, J.; Barek, J.; Fogg, A. G.; Moreira, J. C.; Zima, J. High-performance liquid chromatography of nitrated polycyclic aromatic hydrocarbons. *Analyst.* **1998**, *123*, 9R-18R.

(134) Crimmins, B. S.; Baker, J. E. Improved GC/MS methods for measuring hourly PAH and nitro-PAH concentrations in urban particulate matter. *Atmospheric Environment.* **2006**, *40*, 6764-6779.

(135) Mirivel, G.; Riffault, V.; Galloo, J. C. Simultaneous determination by ultra-performance liquid chromatography-atmospheric pressure chemical ionization time-of-flight mass spectrometry of nitrated and oxygenated PAHs found in air and soot particles. *Analytical and Bioanalytical Chemistry.* **2010**, *397*, 243-256.

(136) Bezabeh, D. Z.; Bamford, H. A.; Schantz, M. M.; Wise, S. A. Determination of nitrated polycyclic aromatic hydrocarbons in diesel particulate-related standard reference materials by using gas chromatography/mass spectrometry with negative ion chemical ionization. *Analytical and Bioanalytical Chemistry.* **2003**, *375*, 381-388.

(137) Marino, F.; Cecinato, A.; Siskos, P. A. Nitro-PAH in ambient particulate matter in the atmosphere of Athens. *Chemosphere.* **2000**, *40*, 533-537.

(138) Kawanaka, Y.; Sakamoto, K.; Wang, N.; Yun, S. J. Simple and sensitive method for determination of nitrated polycyclic aromatic hydrocarbons in diesel exhaust particles by gas chromatography-negative ion chemical ionisation tandem mass spectrometry. *Journal of Chromatography A.* **2007**, *1163*, 312-317.

(139) Glish, G. L.; Vachet, R. W. The basics of mass spectrometry in the twenty-first century. *Nature Reviews Drug Discovery.* **2003**, *2*, 140-150.

(140) Poster, D. L.; Schantz, M. M.; Sander, L. C.; Wise, S. A. Analysis of polycyclic aromatic hydrocarbons (PAHs) in environmental samples: a critical review of gas chromatographic (GC) methods. *Analytical and Bioanalytical Chemistry.* **2006**, *386*, 859-881.

(141) Chen, X.; Li, M. J.; Yi, C. Q.; Tao, Y.; Wang, X. R. Electrochemiluminescence determination of nitro-polycyclic aromatic hydrocarbons using HPLC separation. *Chromatographia.* **2003**, *58*, 571-577. (148) Chuang, J. C.; Pollard, M. A.; Chou, Y. L.; Menton, R. G.; Wilson, N. K. Evaluation of enzyme-linked immunosorbent assay for the determination of polycyclic aromatic hydrocarbons in house dust and residential soil. *Science of the Total Environment.* **1998**, *224*, 189-199.

(142) Kozin, I. S.; Gooijer, C.; Velthorst, N. H. Shpol'skii spectroscopy as a tool in environmental analysis for amino- and nitro-substituted polycyclic aromatic hydrocarbons: A critical evaluation. *Analytica Chimica Acta.* **1996**, *333*, 193-204.

(143) Mirivel, G.; Riffault, V.; Galloo, J.-C. Simultaneous determination by ultra-performance liquid chromatography-atmospheric pressure chemical ionization time-of-flight mass spectrometry of nitrated and oxygenated PAHs found in air and soot particles. *Analytical and Bioanalytical Chemistry. 397.*

(144) Barreto, R. P.; Albuquerque, F. C.; Netto, A. D. P. Optimization of an improved analytical method for the determination of 1-nitropyrene in milligram diesel soot samples by high-performance liquid chromatography-mass spectrometry. *Journal of Chromatography A.* **2007**, *1163*, 219-227.

(145) Scheepers, P. T. J.; Fijneman, P. H. S.; Beenakkers, M. F. M.; Delepper, A.; Thuis, H.; Stevens, D.; Vanrooij, J. G. M.; Noordhoek, J.; Bos, R. P. Immunochemical detection of metabolites of parent and nitro polycyclic aromatic-hydrocarbons in urine samples from persons occupationally exposed to diesel exhaust. *Fresenius J. Anal. Chem.* **1995**, *351*, 660-669.

(146) Zühlke, J.; Knopp, D.; Niessner, R. Determination of 1-nitropyrene with enzyme-linked immunosorbent assay versus high-performance column switching technique. *J. Chromatogr. A.* **1998**, *807*, 209-217.

(147) Fröschl, B.; Knopp, D.; Niessner, R. Possibilities and limitations of an enzyme-linked immunosorbent assay for the detection of 1-nitropyrene in air particulate matter. *Fresenius' Journal of Analytical Chemistry.* **1998**, *360*, 689-692.

(149) Krahl, J.; Knothe, G.; Munack, A.; Ruschel, Y.; Schroder, O.; Hallier, E.; Westphal, G.; Bunger, J. Comparison of exhaust emissions and their mutagenicity from the combustion of biodiesel, vegetable oil, gas-to-liquid and petrodiesel fuels. *Fuel.* **2009**, *88*, 1064-1069.

(150) Vyas, A. P.; Verma, J. L.; Subrahmanyam, N. A review on FAME production processes. *Fuel.* **2010**, *89*, 1-9.

(151) Sivakumar, G.; Vail, D. R.; Xu, J.; Burner, D. M.; Jr., J. O. L.; Ge, X.; Weathers, P. J. Bioethanol and biodiesel: Alternative liquid fuels for future generations. *Energy Life Science.* **2010**, *1*, 8–18.

(152) Pinzi, S.; Garcia, I. L.; Lopez-Gimenez, F. J.; de Castro, M. D. L.; Dorado, G.; Dorado, M. P. The ideal vegetable oil-based biodiesel composition: a review of social, economical and technical implications. *Energy Fuels.* **2009**, *23*, 2325-2341.

(153) Karaosmanoglu, F.; Cigizoglu, K. B.; Tuter, M.; Ertekin, S. Investigation of the refining step of biodiesel production. *Energy Fuels.* **1996**, *10*, 890-895.

(154) Majer, S.; Mueller-Langer, F.; Zeller, V.; Kaltschmitt, M. Implications of biodiesel production and utilisation on global climate - A literature review. *European Journal of Lipid Science and Technology.* **2009**, *111*, 747-762.

(155) Mata, T. M.; Martins, A. A.; Caetano, N. S. Microalgae for biodiesel production and other applications: A review. *Renew. Sust. Energ. Rev.* **2010**, *14*, 217-232.

(156) Shi, A. P.; Wang, Y. B.; Zhu, N.; Ye, L. H.; Zhang, Y. L. Biodiesel production from waste food oil by means of ultrasonic energy and combustion properties as engine fuel. *Information.* **2010**, *13*, 217-227.

(157) Agarwal, A. K. Biofuels (alcohols and biodiesel) applications as fuels for internal combustion engines. *Prog. Energy Combust. Sci.* **2007**, *33*, 233-271.

(158) Fontaras, G.; Kousoulidou, M.; Karavalakis, G.; Tzamkiozis, T.; Pistikopoulos, P.; Ntziachristos, L.; Bakeas, E.; Stournas, S.; Samaras, Z. Effects of low concentration biodiesel blend application on modern passenger cars. Part 1: feedstock impact on regulated pollutants, fuel consumption and particle emissions. *Environ Pollut.* **2010**, *158*, 1451-1460.

(159) Qi, D. H.; Chen, H.; Matthews, R. D.; Bian, Y. Z. H. Combustion and emission characteristics of ethanol-biodiesel-water micro-emulsions used in a direct injection compression ignition engine. *Fuel.* **2010**, *89*, 958-964.

(160) Graboski, M. S.; McCormick, R. L. Combustion of fat and vegetable oil derived fuels in diesel engines. *Prog. Energy Combust. Sci.* **1998**, *24*, 125-164.

(161) Karavalakis, G.; Stournas, S.; Bakeas, E. Light vehicle regulated and unregulated emissions from different biodiesels. *Sci. Total Environ.* **2009**, *407*, 3338-3346.

(162) Nelson, G. O. *Controlled test atmospheres*; Ann Arbor Science Publishers: Michigan, U.S.A, 1971.

(163) Wu, C. W.; Chen, R. H.; Pu, J. Y.; Lin, T. H. The influence of air-fuel ratio on engine performance and pollutant emission of an SI engine using ethanol-gasoline-blended fuels. *Atmos. Environ.* **2004**, *38*, 7093-7100.

(164) Jia, L. W.; Shen, M. Q.; Wang, J.; Lin, M. Q. Influence of ethanol-gasoline blended fuel on emission characteristics from a four-stroke motorcycle engine. *J. Hazard. Mater.* **2005**, *123*, 29-34.

(165) Borras, E.; Tortajada-Genaro, L. A.; Vazquez, M.; Zielinska, B. Polycyclic aromatic hydrocarbon exhaust emissions from different reformulated diesel fuels and engine operating conditions. *Atmos. Environ.* **2009**, *43*, 5944-5952.

(166) Ballesteros, R.; Hernandez, J. J.; Lyons, L. L. An experimental study of the influence of biofuel origin on particle-associated PAH emissions. *Atmos. Environ.* **2010**, *44*, 930-938.

(167) Karavalakis, G.; Stournas, S.; Bakeas, E. Effects of diesel/biodiesel blends on regulated and unregulated pollutants from a passenger vehicle operated over the European and the Athens driving cycles. *Atmos. Environ.* **2009**, *43*, 1745-1752.

(168) Durbin, T. D.; Collins, J. R.; Norbeck, J. M.; Smith, M. R. Effects of biodiesel, biodiesel blends, and a synthetic diesel on emissions from light heavy-duty diesel vehicles. *Environ. Sci. Technol.* **2000**, *34*, 349-355.

(169) Turrio-Baldassarri, L.; Battistelli, C. L.; Conti, L.; Crebelli, R.; De Berardis, B.; Iamiceli, A. L.; Gambino, M.; Iannaccone, S. Emission comparison of urban bus engine fueled with diesel oil and 'biodiesel' blend. *Sci. Total Environ.* **2004**, *327*, 147-162.

(170) Karavalakis, G.; Fontaras, G.; Ampatzoglou, D.; Kousoulidou, M.; Stournas, S.; Samaras, Z.; Bakeas, E. Effects of low concentration biodiesel blends application on modern passenger cars. Part 3: impact on PAH, nitro-PAH, and oxy-PAH emissions. *Environ Pollut.* **2010**, *158*, 1584-1594.

(171) Guarieiro, L. L. N.; Pereira, P. A. D.; Torres, E. A.; da Rocha, G. O.; de Andrade, J. B. Carbonyl compounds emitted by a diesel engine fuelled with diesel and biodiesel-diesel blends: Sampling optimization and emissions profile. *Atmos. Environ.* **2008**, *42*, 8211-8218.

(172) Brito, J. M.; Belotti, L.; Toledo, A. C.; Antonangelo, L.; Silva, F. S.; Alvim, D. S.; Andre, P. A.; Saldiva, P. H. N.; Rivero, D. Acute cardiovascular and inflammatory toxicity induced by inhalation of diesel and biodiesel exhaust particles. *Toxicological Sciences.* **2010**, *116*, 67-78.

(173) Buenger, J.; Krahl, J.; Munack, A.; Ruschel, Y.; Schroeder, O.; Emmert, B.; West, G.; Mueller, M.; Hallier, E.; Bruening, T. Strong mutagenic effects of diesel engine emissions using vegetable oil as fuel. *Archives of Toxicology.* **2007**, *81*, 599-603.

(174) Helsper, C.; Moelter, W.; Loeffler, F.; Wadenpohl, C.; Kaufmann, S.; Wenninger, G. Investigations of a new aerosol generator for the production of carbon aggregate particles. *Atmospheric Environment.* **1993**, *27A*, 1271-1275.

(175) Niessner, R. Coated particles: preliminary results of laboratory studies on interaction of ammonia with coated sulfuric acid droplets or hydrogen sulfate particles. *Science of the Total Environment.* **1984**, *36*, 353-362.

(176) Niessner, R. The chemical response of the photo-electric aerosol sensor (PAS) to different aerosol systems. *Journal of Aerosol Science.* **1986**, *17*, 705-714.

(177) *Toxicological profile for polycyclic aromatic hydrocarbons*. US Department of Health & Human Services, Agency for Toxic Substances and Disease Registry, Washington, DC, August, 1985

(178) Carrara, M. *Experimental Study of the oxidation readiness of soot by Raman microscopy, thermogravimetric oxidation analysis and FTIR spectroscopy*; Master-Thesis, Technische Universität München, 2007.

(179) Teckentrup, A.; Klockow, D. Preparation of refillable permeation tubes. *Analytical Chemistry.* **1978**, *50*, 1728.

(180) McMurry, P. H. A review of atmospheric aerosol measurements. *Atmospheric Environment.* **2000**, *34*, 1959-1999.

(181) TSI *Instruction Manual, 3071 Aerosol Classifier*. TSI incorporated, 1985.

(182) Peters, T. M.; Chein, H. M.; Lundgren, D. A.; Keady, P. B. Comparison and combination of aerosol-size distributions measured with a low-pressure impactor, differential mobility particle sizer, electrical aerosol analyzer, and aerodynamic particle sizer. *Aerosol Science and Technology.* **1993**, *19*, 396-405.

(183) TSI *Instruction manual, 3025 CPC*. TSI incorporated, 1985.

(184) Saltzman, B. E. Colorimetric microdetermination of nitrogen dioxide in the atmosphere. *Anal. Chem.* . **1954**, *26*, 1949-1955.

(185) Kaneko, K.; Imai, J. Adsorption of NO_2 on activated carbon-fibers. *Carbon.* **1989**, *27*, 954-955.

(186) Rubel, A. M.; Stewart, M. L.; Stencel, J. M. Activated carbon for control of nitrogen oxide emissions. *J. Mater. Res.* **1995**, *10*, 562-567.

(187) Neathery, J. K.; Rubel, A. M.; Stencel, J. M. Uptake of NO_x by activated carbons: bench-scale and pilot-plant testing. *Carbon.* **1997**, *35*, 1321-1327.

(188) Rubel, A. M.; Stencel, J. M. Effect of pressure on NO_x adsorption by activated carbons. *Energy Fuels.* **1996**, *10*, 704-708.

(189) Mastral, A. M.; Garcia, T.; Murillo, R.; Callen, M. S.; Lopez, J. M.; Navarro, M. V. Measurements of polycyclic aromatic hydrocarbon adsorption on activated carbons at very low concentrations. *Industrial & Engineering Chemistry Research.* **2003**, *42*, 155-161.

(190) Zhou, H. C.; Zhong, Z. P.; Jin, B. S.; Huang, Y. H.; Xiao, R. Experimental study on the removal of PAHs using in-duct activated carbon injection. *Chemosphere.* **2005**, *59*, 861-869.

(191) Possanzini, M.; Febo, A.; Cecchini, F. Development of a potassium iodide annular denuder for nitrogen oxide collection. *Analytical Letters.* **1984**, *17*, 887-896.

(192) Mueller, J. O.; Su, D. S.; Jentoft, R. E.; Wild, U.; Schloegl, R. Diesel engine exhaust emission: oxidative behavior and microstructure of black smoke soot particulate. *Environmental Science and Technology.* **2006**, *40*, 1231-1236.

(193) Mueller, J. O.; Su, D. S.; Jentoft, R. E.; Kroehnert, J.; Jentoft, F. C.; Schloegl, R. Morphology-controlled reactivity of carbonaceous materials towards oxidation. *Catalysis Today.* **2005**, *102-103*, 259-265.

(194) Messerer, A.; Schmid, H.-J.; Knab, C.; Poeschl, U.; Niessner, R. Enhancement of the deposition of ultrafine diesel soot particles by microsphere coating on metal support-based catalyst structures. *Chemie Ingenieur Technik.* **2004**, *76*, 1092-1096.

(195) *Official Journal of the European Communities L 286*, 1999.

(196) Dyker, G.; Kadzimirsz, D.; Thone, A. Synthesis of 1-and 3-nitrobenzo[a]pyrene. *European Journal of Organic Chemistry.* **2003**, 3162-3166.

(197) Delhomme, O.; Herckes, P.; Millet, M. Determination of nitro-polycyclic aromatic hydrocarbons in atmospheric aerosols using HPLC fluorescence with a post-column derivatisation technique. *Analytical and Bioanalytical Chemistry.* **2007**, *389*, 1953-1959.

(198) Dickhut, R. M.; Miller, K. E.; Andren, A. W. Evaluation of total molecular surface area for predicting air-water partitioning properties of hydrophobic aromatic chemicals. *Chemosphere.* **1994**, *29*, 283-297.

(199) Wang, H. M.; Hasegawa, K.; Kagaya, S. Nitration of pyrene adsorbed on silica particles by nitrogen dioxide under simulated atmospheric conditions. *Chemosphere.* **1999**, *39*, 1923-1936.

(200) Ramdahl, T.; Bjorseth, A.; Lokensgard, D. M.; Pitts, J. N. Nitration of polycyclic aromatic-hydrocarbons adsorbed to different carriers in a fluidized-bed reactor. *Chemosphere.* **1984**, *13*, 527-534.

(201) Miet, K.; Le Menach, K.; Flaud, P. M.; Budzinski, H.; Villenave, E. Heterogeneous reactivity of pyrene and 1-nitropyrene with NO_2: Kinetics, product yields and mechanism. *Atmospheric Environment.* **2009**, *43*, 837-843.

(202) Kristovich, R. L.; Dutta, P. K. Nitration of benzo[a]pyrene adsorbed on coal fly ash particles by nitrogen dioxide: Role of thermal activation. *Environmental Science & Technology.* **2005**, *39*, 6971-6977.

(203) Wang, H. M.; Hasegawa, K.; Kagaya, S. The nitration of pyrene adsorbed on silica particles by nitrogen dioxide. *Chemosphere.* **2000**, *41*, 1479-1484.

(204) Lammel, G.; Perner, D. The atmospheric aerosol as a source of nitrous acid in the polluted atmosphere. *Journal of Aerosol Science.* **1988**, *19*, 1199-1202.

(205) Bosch, E.; Kochi, J. K. Thermal and photochemical nitration of aromatic hydrocarbons with nitrogen dioxide. *Journal of Organic Chemistry.* **1994**, *59*, 3314-3325.

(206) Kalberer, M.; Ammann, M.; Arens, F.; Gaggeler, H. W.; Baltensperger, U. Heterogeneous formation of nitrous acid (HONO) on soot aerosol particles. *J. Geophys. Res.-Atmos.* **1999**, *104*, 13825-13832.

(207) de Queiroz, J. F.; Carneiro, J. W. D.; Sabino, A. A.; Sparrapan, R.; Eberlin, M. N.; Esteves, P. M. Electrophilic aromatic nitration: Understanding its mechanism and substituent effects. *Journal of Organic Chemistry.* **2006**, *71*, 6192-6203.

(208) Ghigo, G.; Causa, M.; Maranzana, A.; Tonachini, G. Aromatic hydrocarbon nitration under tropospheric and combustion conditions. A theoretical mechanistic study. *Journal of Physical Chemistry A.* **2006**, *110*, 13270-13282.

(209) Squadrito, G. L.; Fronczek, F. R.; Church, D. F.; Pryor, W. A. Free-radical nitration of naphthalene with nitrogen dioxide in CCl_4 and implications for environmental nitrations. *Journal of Organic Chemistry.* **1989**, *54*, 548-552.

(210) Shiraiwa, M.; Garland, R. M.; Poschl, U. Kinetic double-layer model of aerosol surface chemistry and gas-particle interactions (K2-SURF): Degradation of polycyclic aromatic hydrocarbons exposed to O-3, NO2, H2O, OH and NO3. *Atmospheric Chemistry and Physics.* **2009**, *9*, 9571-9586.

(211) He, C.; Ge, Y. S.; Tan, J. W.; You, K. W.; Han, X. K.; Wang, J. F. Characteristics of polycyclic aromatic hydrocarbons emissions of diesel engine fueled with biodiesel and diesel. *Fuel.* **2010**, *89*, 2040-2046.

(212) Zielinska, B.; Sagebiel, J.; Arnott, W. P.; Rogers, C. F.; Kelly, K. E.; Wagner, D. A.; Lighty, J. S.; Sarofim, A. F.; Palmer, G. Phase and size distribution of polycyclic aromatic hydrocarbons in diesel and gasoline vehicle emissions. *Environmental Science & Technology.* **2004**, *38*, 2557-2567.

(213) Pitts, J. N. Nitration of gaseous polycyclic aromatic-hydrocarbons in simulated and ambient urban atmospheres - a source of mutagenic nitroarenes. *Atmospheric Environment.* **1987**, *21*, 2531-2547.

MIX
Papier aus verantwortungsvollen Quellen
Paper from responsible sources
FSC® C105338

Printed by Books on Demand GmbH, Norderstedt / Germany